PhotoManual&
DissectionGuide**of**the

FETAL PIG

with**Sheep**Heart**Brain**Eye

Fred**Bohensky**

SQUAREONE
EDUCATION GUIDES

SQUARE ONE'S ANATOMY SERIES

Photo Manual and Dissection Guide of the Cat
Fred Bohensky

Photo Manual and Dissection Guide of the Fetal Pig
Fred Bohensky

Photo Manual and Dissection Guide of the Frog
Fred Bohensky

Photo Manual and Dissection Guide of the Rat
Fred Bohensky

Photo Manual and Dissection Guide of the Shark
Fred Bohensky

ISBN 0-7570-0030-4   (Fetal Pig)

Printed in the United States of America

CONTENTS

DEDICATION

To my daughter C. R., who took time from a very busy schedule to type the manuscript. Often she was heard pounding the keys till the early hours of the morning.

To my wife Esther, who painstakingly proofread the original and final versions.

Their help and constant encouragement helped to see this project to its successful completion.

PREFACE

"A picture is worth a thousand words." With this adage in mind, full page 8" x 10" fully labeled photographs of the fetal pig dissection are presented. These, together with the many diagrams, will help the Biology student to better visualize the parts and structures he is to find on his own specimen.

Comparisons to the anatomy and physiology of man are made throughout. In most cases, the interest of the student, as well as the focus of the course, is upon man. The fetal pig is used because of its relative abundance, low cost, ease of storage, as well as its similarity to humans. The fetal pig continues to be a popular laboratory specimen in courses of General Biology, Anatomy and Physiology, Comparative Anatomy, as well as in Mammalian Anatomy.

This manual is intended as a tool for the student. The author has, after many years experience, decided to present model dissections, with clear concise instructions. These are presented in the large, clearly labeled photographs as well as in the accompanying text.

Even the second-year Biology student is often overwhelmed when he or she is handed a dissection specimen at the start of the semester, and instructed to follow the procedures outlined in the lab manual. Difficulty is encountered in relating the structures seen in the actual specimen with the lab manual's diagrams, which are often poorly drawn, small, or inaccurately labeled. To the untrained eye, nerves, blood vessels, and the smaller structures look very different from the figures as depicted. These diagrams are generally idealized versions of "perfect" dissections, a far cry from the student's often clumsy attempts.

Some of the structures seen in the photographs are as big as those of the actual specimens, some are bigger. Location and identification of structures are thus facilitated. Valuable time is gained by preventing needless exploratory incisions. It is often difficult for the novice to determine the limits of an organ, where it begins or ends. Various structures have therefore been outlined by lines or a series of dots. The names of others have been printed on the organs themselves. Some photos show the same area from different vantage points. For example, while one photo shows the entire arterial system, the following two show magnified details of the arteries in the upper and lower regions. The brain of the fetal pig is seen from top and side views . Before exams, these photos will vividly bring to mind the actual dissections.

The fetal pigs described and photographed are about a foot long, nearly fully developed prior to birth. They are doubly injected with red latex in the arteries, blue in the veins.

A Self-Quiz for students is included at the end of each unit. it consists of short answers, definitions, and the labeling of photographs. These worksheets may be removed from the book and submitted to the instructor for correction.

The sheep heart, brain, and eye are included. These supplement nicely the much smaller structures of the fetal pig. They enable one to make a more detailed as well as comparative study.

The author wishes to acknowledge the help extended by the Media Production Center of the College of Staten Island. He is particularly grateful for the assistance of Mr. Joseph Rickard whose fine lenswork is seen throughout this manual.

Fred Bohensky
Staten Island, New York

INTRODUCTION TO THE FETAL PIG

For most students, this is their first major dissection. A few words of introduction are in order.

The fetal pigs are unborn animals. In processing the sows (mature female swine) for meat, the *uterine horns* often reveal unborn litters. These are removed and made available for biological study. Note the *umbilical cord* of your specimen. There are usually 7 to 12 young in a single litter.

Determination of the Age of the Fetal Pig

The age of the fetal pig is determined by its length. Measure body length along the back following the natural curvature of the spine from the tip of the snout to the base (the start) of the tail, not to its tip. Use a string for this. The actual length is determined by measuring the string with a centimeter ruler.

Refer to the following table which relates the body length to the number of days of gestation. Convert the *centimeter* values shown in the first column to *millimeters* and enter these in the second column.

LENGTH OF FETUS		APPROXIMATE AGE (days)
cm	**mm**	
1.1		21
1.7		35
2.8		49
4.0		56
22.0		100
30.0		Full term 112-115

The fetal pig assigned to you measures _____ cm (_____ mm) in length. It seems to be approximately _____ days old.

Two students will share one specimen and will be responsible for it till the end of the course.

The Domestic Pig

Domestic hogs belong to the same species as the European wild hog, *Sus scrofa*. They belong to the *Order Artiodactyla* (even-toed hoofed mammals) which includes such varieties as cattle, sheep, goats, deer, camels, giraffes, and hippopotamuses.

Sus scrofa is *omnivorous* like man, feeding both on plant and animal matter. The *body temperature* of an adult pig is slightly higher than that of man. The life span of the pig is 15 to 20 years. There are many biochemical similarities between man and pig (composition of body fluids, specific enzymes, etc.). Pigs are valuable experimental animals for the study of effects of drugs and radiations. A "mini-pig" has been developed for this purpose; it weights 100-150 pounds when fully grown, rather than the 900 pounds reached by some domestic breeds (e.g., Yorkshire).

The Mammals

Since we are about to study the structures and functions of a type of mammal, we ought to consider this most highly developed animal form more closely.

Both the pig and man belong to the class of vertebrates known as Mammalia. Mammals are the most highly developed animal form. While our dissection subject is the pig, we will be making constant reference to man in both the text and in the diagrams presented.

Mammals are a class of *Vertebrates* or backboned animals, that also includes the:

Fish

Amphibians (frog, toads, and salamanders)

Reptiles (lizards, snakes, turtles, and crocodiles)

Birds

Mammals range in size from minute shrews, which weigh only about two grams (0.002 kg) to giant blue whales, which weigh up to 115 tons (115,000 kg).

The two chief mammalian characteristics which set these animals off from the other classes of Vertebrates are:

Skin covered with hair or fur.

Milk-producing glands (mammary glands) in the female to nurse the young.

The most primitive living mammals belong to the Order *Monotremata*. These animals, native to the Australian region, include the duck-billed platypus and the spiny anteater. They lay a reptilian type of egg. When hatched, the young receive nourishment from the mother's mammary glands.

A more advanced form of mammal, belonging to the order of the pouched mammals, the *Marsupialia,* gives birth to live young (viviparous). However, the young are born at a very early stage of development and continue their maturation in a pouch, where they attach to nipples of the mammary glands. Australian kangaroos and American opossums are members of this order.

The most familiar mammals belong to the subclass *Eutheria,* or *placental* mammals. They include the pig as well as man. Other members of this diverse group include the dogs, cats, cattle, rats, whales, lions, tigers, apes, monkeys, giraffes, and hippopotamuses, and many others.

Their embryos develop within the *uterus* (womb) and are nourished by a special structure, the *placenta,* until they emerge highly developed. Most can walk and even run within a few days of birth; man, however, is helpless for the longest period of time.

During *gestation* (the period during which the embryo develops within the uterus), there is an exchange of substances between the blood of the mother and the embryo across the placenta through the *umbilical cord.*

Students in the Health Sciences, please note that although your primary interest is the human organism, most organs and tissues of the pig are structurally and functionally similar to those of man. Even their names are quite similar, most often identical. Charts, models, and skeletons of human anatomy should be made available to you while you are studying comparable features of the pig.

ANATOMICAL TERMINOLOGY

Some basic biological terminology should be studied at this time. Familiarize yourself with the following words and learn to use them in referring to the location of the body parts of your specimen.

Directions or Positions

Anterior (Cranial)	— toward the head
Posterior (Caudal)	— toward the tail
Dorsal (Superior)	— toward the backbone
Ventral (Inferior)	— toward the belly
Lateral	— toward the side
Medial	— toward the midline
Proximal	— lying near the point of reference
Distal	— lying further from the point of reference

Note: The terms in parentheses are synonymous only when referring to a quadruped such as a pig. In man these terms have different meanings (see diagrams at the end of this section).

Planes or Sections Through the Body

Transverse (Cross Section)	— perpendicular to the long axis of the body
Sagittal	— a longitudinal section separating the body into right and left sides
Frontal (Coronal)	— a longitudinal section dividing the specimen into dorsal and ventral parts

In man, the *anterior* and *ventral* surfaces are identical; both terms refer to a person's front or belly side. Similarly, the terms *posterior* and *dorsal* are identical, referring to a person's back surface, the area near the spinal cord.

In the pig and other four-legged animals these terms are not at all identical. *Ventral* still refers to the belly portion, but the pig's belly is not at its *anterior* or head end, but on the lower (*inferior*) surface. *Dorsal* still refers to the area of the spinal cord, but the rat's spine is not located along its *posterior* or tail end, but on its upper (*superior*) surface.

Other terms indicating position or direction will appear in the text. For example, the terms *superficial* (or external) and *deep* (or internal) will be used when describing muscles. The terms *cranial* and *caudal* will indicate the head and tail ends respectively.

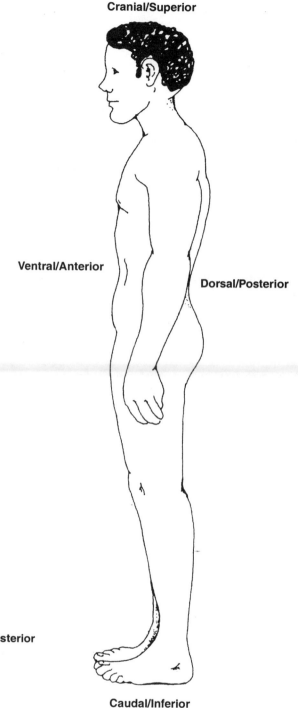

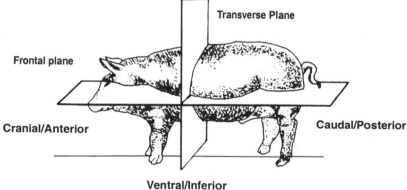

DIRECTIONAL TERMS for the Pig and Man

EXTERNAL FEATURES

Preliminary Procedures

When you obtain your specimen remove it carefully from the plastic bag container and place into a dissection pan. Do not discard the preservative fluid in the bag. The animal will be returned to the bag at the end of each lab session and the fluid will keep the specimen in good moist condition.

Obtain an identification tag and a rubber band for your specimen bag. At the close of each session, after the pig has been returned to the bag, twist the top of the bag and seal tightly with the rubber band. Then, attach the tag. Write your name, your partner's name, the section number, and the instructor's name, on the label.

Examination

Examine the specimen you have received. Lay the pig on its side as in the first photo.
The body is divided into four readily identifiable areas:
- head (cranial)
- neck (cervical)
- trunk (thoracic and abdominal)
- tail (caudal)

Appendages (Limbs)

The pig is a *quadruped*, in contrast to man who is a *biped*. This refers to four- and two-legged locomotion, respectively. The pig walks on the toes; this is called *digitigrade* locomotion. Man walks on the sole of the foot; this is called *plantigrade* locomotion.

In the photo note the positions of the elbow and wrist, ankle and knee. Count the digits (toes) on each foot. Each foot has four toes. The middle two are flattened and have hooves.

Head

Locate the following parts of the head:

Snout — The snout of the pig has a blunt tip, ending with a disc-like, pliable but firm structure composed of *cartilage*. The tip of the nose is also strengthened by bone. This permits the pig to use the snout to push, lift weights and dig.

External Nares — These are nostrils opening in the cartilaginous disc of the snout. They open into the nasal cavity. Here the inhaled air is warmed, filtered and humidified.

Pinnae — These are the external ears. They are also composed of cartilage, just as the human ear. The *external auditory meatus* is the opening for the *external auditory canal* which leads to the *tympanic membrane* (eardrum), and to the *middle ear*.

Eyes — Spread the upper and lower eyelids. In the inner corner of the eye locate a third lid-like structure, the *nictitating membrane*. Does the pig have eyelashes?

Trunk

Locate the following parts of the trunk:

Umbilical Cord — This structure more than any other, identifies the animal as a fetus. It extends from the mid-ventral abdominal surface to the *placenta*. It functions in the procurement of food and oxygen for the fetus from the mother, and the movement of wastes from fetus to mother.

Use your scissors to cut the umbilical cord about a half inch from the abdomen. Observe the two red *umbilical arteries* and the much larger blue *umbilical vein* running through the cord. A smaller *allantoic duct* will also be found.

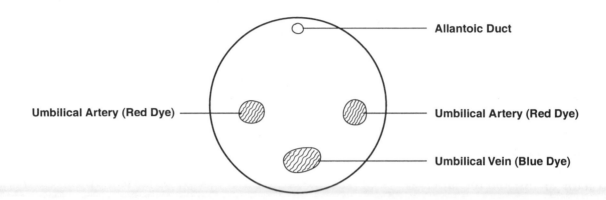

UMBILICAL CORD (Cross Section)

Anus — This is the terminal opening of the digestive tract. it is located just ventral to the base of the tail in both males and females. Simply lift the tail to find the anus.

Urogenital openings and *mammary papillae will* be described in the next section. Note the paper-thin covering upon the fetal pig's entire body, the *periderm*. This may easily be peeled off.

At the end of each dissection session, replace the pig in the plastic bag. Add an ounce or two of preservative fluid. Twist the top and seal securely with a rubber band. This prevents your specimen from drying out between dissection sessions. Attach the identification tag to the outside of the bag.

Note the large incision on the pig's neck in the photo (p. 8). This was made at the time the colored latex was injected into the pig's blood vessel.

In the same photo you can also observe how the fetal pig is positioned for most of the following dissection sessions. He is tied down in the dissection pan, ventral surface upward. The string extends from one limb across to the other, passing beneath the bottom of the pan. A slip knot facilitates the tightening and loosening of the string as needed during the dissection.,

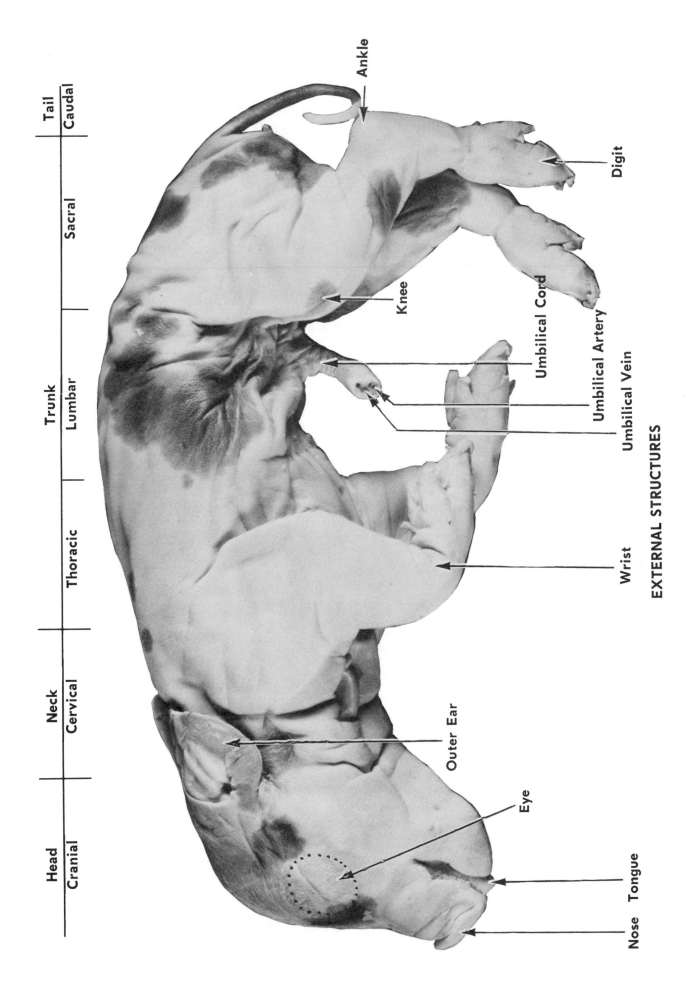

EXTERNAL STRUCTURES

Head | Neck | Trunk | Tail
Cranial | Cervical | Thoracic | Lumbar | Sacral | Caudal

Tail
Ankle
Digit
Knee
Umbilical Cord
Umbilical Artery
Umbilical Vein
Wrist
Outer Ear
Eye
Nose   Tongue

7

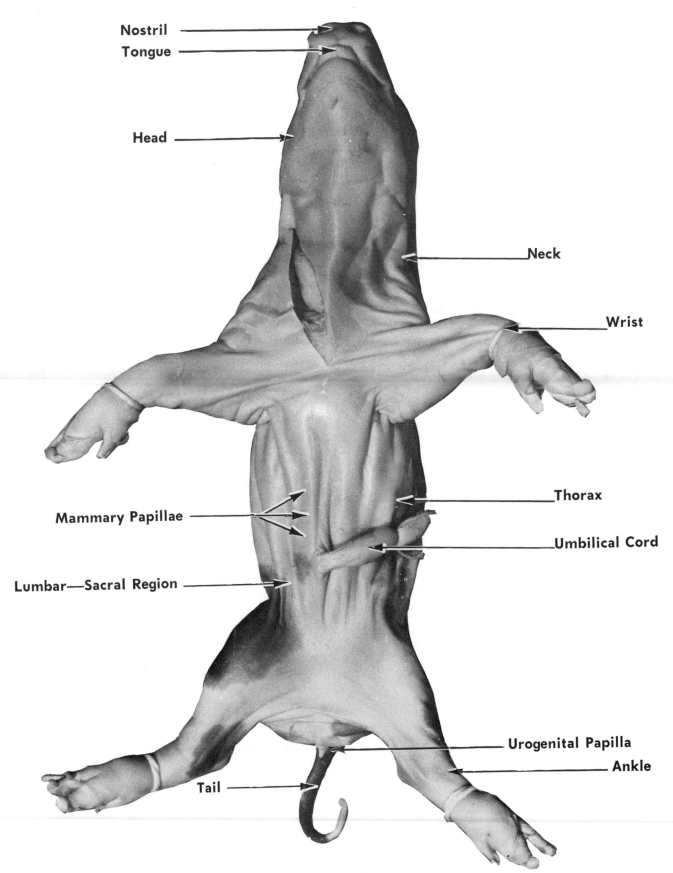

Nostril

Tongue

Head

Neck

Wrist

Thorax

Mammary Papillae

Umbilical Cord

Lumbar—Sacral Region

Urogenital Papilla

Ankle

Tail

EXTERNAL STRUCTURES (Ventral View)

MALE AND FEMALE

Although the animals are as yet unborn, differences between the sexes are readily seen. The older the specimen the more pronounced these external differences will be.

Female: (Symbol ♀)

The female is identified by the *urogenital papilla*. This is a small fleshy conical prejection ventral to the anus. (The anus is ventral to the tail and is clearly seen in both males and females, when the tail is lifted.) Locate the female's *external genital opening* at the base of the urogenital papilla. As the term urogenital indicates, this is the external opening for both urinary wastes and the reproductive or genital system.

Male: (Symbol ♂)

The male's *testes* (testis singular) lie in the *scrotum,* a double pouch structure ventral to the tail. In older specimens this area is enlarged and readily visible. In younger animals it may be necessary to touch the area to detect the testes.

The *urogenital opening* in males is located on the mid-ventral surface, posterior to (below) the umbilical cord. It is the opening of the *penis.*

The *penis* is internal, but may be detected under the skin by pressing with your finger tip (palpating) along the mid-ventral surface, between the urogenital opening and the scrotum.

Males and Females:

Both males and females possess *mammary papillae.* In mature females these become the nipples by which the young receive the milk from the *mammary glands*. In males and in these fetal animals, the glandular milk producing structures are not developed. However, the mammary papillae are present in all specimens.

How many pairs of mammary papillae do you count in your specimen? Do all specimens in the class possess the same number?

Examine the fetal pigs of other students to determine the sex of the specimens.

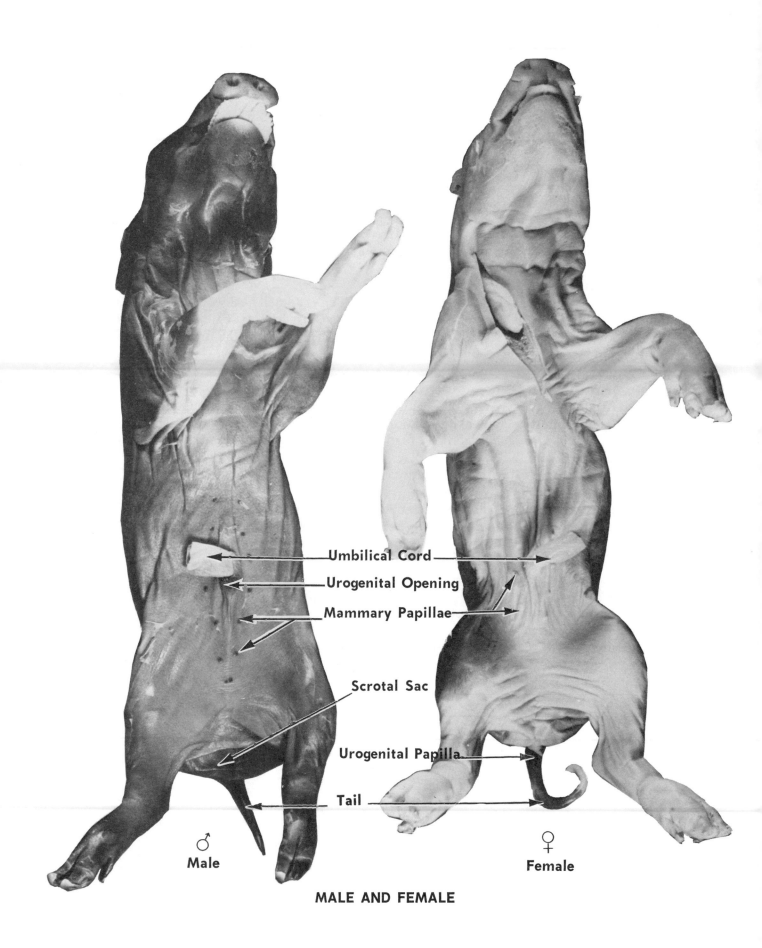

Umbilical Cord

Urogenital Opening

Mammary Papillae

Scrotal Sac

Urogenital Papilla

Tail

♂
Male

♀
Female

MALE AND FEMALE

THE SKELETON

The skeleton of the fetal pig has not yet fully *ossified*, or hardened, to bone. Much of the skeleton is still composed of cartilage. It is therefore best to obtain an actual mounted skeleton for study. It is also possible to obtain adult bones or joints of adult pigs from a local butcher shop. In addition, diagrams of the adult pig skeleton are here included. Study these carefully.

The human skeleton, fully mounted, should also be provided. A fully labeled diagram is included. Throughout the study of the pig, compare its structures to those of man. You will note that the bones are named identically. Generally, it is only the number of bones that differ. Their arrangement in the various mammals is determined by differences in the method of support and locomotion.

Feet

Observe the position of the *feet* in the diagram of the adult pig. While man walks on the sole of the foot: *plantigrade* locomotion, the pig walks on his toes: *digitigrade* locomotion.

The pig belongs to the order Artiodactyla, or even-toed hoofed (ungulate) mammals. Note that the feet are narrow and the foot bones are separate, not fused. The first digit is absent. The middle two are flattened (they are the third and fourth digits) and have hooves. The lateral (side) toes represent the second and fifth digits.

Raise your fetal pig to the walking position. Orient one of the feet of the pig in the walking position. Which toes touch the ground? Explain by noting the position of the foot and the digit number.

Teeth

An animal's diet is revealed by its *dentition* pattern. This refers to the types of teeth the animal possesses, their number, and arrangement.

Sharp and pointy *incisors, canines,* and *pre-molars* predominate in *carnivorous* animals such as the dog, cat, tiger, and others. *Herbivorous* animals such as horses and cows, possess incisors for shearing grass and other vegetable matter. These are followed by rows of large flattened *molars* with broad grinding surfaces toward the rear of the mouth.

The pig and human are *omnivorous*, this diet consisting of both animal and plant matter. It combines sharp pointy incisors with grinding pre-molars and molars.

The dental formulas of the adult pig and man are compared below:

Pig — $\qquad I\dfrac{3}{3}, C\dfrac{1}{1}, P\dfrac{4}{4}, M\dfrac{3}{4}$

Man — $\qquad I\dfrac{2}{2}, C\dfrac{1}{1}, P\dfrac{2}{2}, M\dfrac{3}{3}$

The letters refer to the types of teeth:
I — Incisor,   C — Canine,   P — Premolar,   M — Molar

The upper set of numbers refers to the number of teeth in half of the upper jaw, the lower set of numbers refers to the number of teeth in half of the lower jaw. Thus the total number of teeth in adult pigs is 44 and 32 in man.

The skeleton of all vertebrates is internal, known as the *endoskeleton*. It may be divided into two main areas.

Axial skeleton — includes the bones of the main longitudinal axis of the body.

Appendicular skeleton — the bones of the appendages plus their supporting girdles.

Axial Skeleton

The axial skeleton is composed of the:
- Skull
- Vertebral Column
- Ribs
- Sternum

Skull — The skull consists of the *cranium*, the bony vault protecting the brain, and the *facial bones*.

Cranium — The cranium is composed of 8 bones: 1 *frontal*, 2 *Parietals*, 2 *temporals*, 1 *occipital*, 1 *sphenoid*, and 1 *ethmoid*.

Facial Bones — The pig has 19 facial bones, man has only 14. The facial bones of the pig are elongated, particularly the *maxilla* and the *nasal bones*. The *premaxilla* bones, between the maxilla and nasal bones, are not found in man. Identify and learn the names of the facial bones: *maxilla, zygomatic, lacrimal, nasal, vomer, palotine,* and *mandible*.

Vertebral Column — The outstanding characteristic of the vertebrates is the possession of a backbone or *vertebral column*. It serves as an attachment for the muscles of the back and its support. The soft delicate *spinal cord* runs through the vertebral bones, and is protected by them. The vertebral column of the pig is composed of 51-56 bones, that of man only 33.

A typical vertebral bone consists of the *body* or *centrum, neural arch* with *spine*, a pair of lateral and transverse *processes*, and the posterior *articular facets* at the point where the vertebra meet. A pad of cartilage, the *intervertebral disc*, forms a protective cushion between adjacent vertebrae.

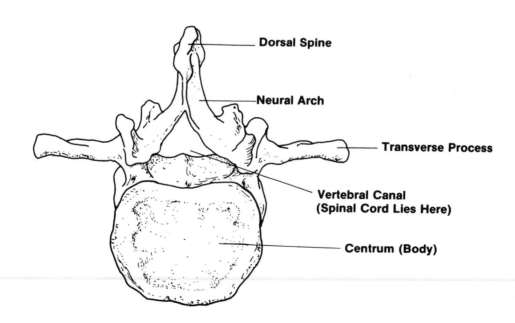

VERTEBRA (Human, Lumbar)

Five sections of the vertebral column are readily identified:

1. **Cervical** — These 7 bones are in the neck region. The topmost two are the *atlas* and the *axis*. They permit free movement and rotation of the head. Virtually all mammals, even tall giraffes, have 7 cervical vertebrae.
2. **Thoracic** — They number 12 in man, 14-15 in the pig. They have prominent dorsal spines and *articular costal* facets for the attachment of ribs.
3. **Lumbar** — These number 5 in man and 6-7 in the pig. These are characterized by massive centra, reduced dorsal spines, and long transverse processes.
4. **Sacral** — In man 5 sacral vertebrae are fused into one bone, the *sacrum*. The pig has only 4.
5. **Caudal** — There are 20-23 caudal vertebrae in the pig extending out into the tail. In man 3-5 bones are fused to form the *coccyx*.

Variations in the number of vertebrae:

Region	Pig	Cat	Man
Cervical	7	7	7
Thoracic	14-15	13	12
Lumbar	6-7	7	5
Sacral	4	3	5
Caudal	20-23	21-25	3-5 (Coccyx)

Ribs — In man there are 12 pairs of ribs. Of these, the upper 7 pairs are known as "true" ribs. They articulate with the thoracic vertebrae as well as with the *Sternum*, or breastbone. The attachment to the sternum is by way of *costal cartilage*, not bone. The next 3 are called "false" ribs. They do not articulate with the sternum directly only by way of the seventh rib. The last two pairs of ribs are called "floating" ribs. They articulate with the thoracic vertebrae, but do not reach the sternum at all.

In the pig there are 14-15 pairs. Seven pairs articulate directly with the sternum by way of a short costal cartilage. The other ribs are joined together by their costal cartilage before attaching to the sternum.

Sternum — The sternum of pig and man are very similar. It consists of the anterior portion, the *manubrium*; a central portion, the *gladiolus*; and a posterior portion made of cartilage, the *xiphoid process*.

The *hyoid* bone, at the top of the trachea, and the 3 pairs of *auditory ossicles* in the middle ear: the *hammer, anvil,* and *stirrup,* are generally considered part of the axial skeleton, although embryologically they have quite different origins and are traced to the gill apparatus, jaw, and pharynx of fish.

Appendicular Skeleton

The appendicular skeleton consists of the *pectoral girdle* and the attached *forelimbs*, and the *pelvic girdle* and the attached *hind limbs*.

Pectoral Girdle — In man there are two pairs of bones which comprise this structure, the two *scapulas* and two *clavicles*. The pig does not possess clavicles. Much of the pig's scapula remains cartilaginous even in the adult.

Forelimbs — In man each forelimb is composed of the *humerus* (upper arm), *ulna* and *radius* (lower arm), 8 *carpal* bones (wrist), 5 *metacarpals* (palm), and 5 *digits* composed of 14 *phalanges*.

In the pig the humerus, ulna and radius are reduced. There are 8 carpal bones, 4 metacarpals, and 4 digits each composed of 3 phalanges.

Pelvic Girdle — The pelvic girdle of the pig, as of man, consists primarily of a paired bone, the *innominate*, or *os coxa*. It is, in turn, formed from three separate bones which fused during fetal development. They are, the *ilium* laterally, the *ischium* posteriorly, and the *pubis* ventrally. A pad of cartilage, the *pubic symphysis*, lies at the mid-ventral juncture of the pubic bones and unites them ventrally. A spherical depression, the *acetabulum*, along the ventro-lateral border of the girdle, serves as the point of articulation for the hind limb. It is the "socket" that receives the ball-shaped head of the femur.

Hind Limbs — In man each leg is composed of the *femur* (thigh bone), *Patella* (knee cap) , *tibia* and *fibula* (shin and calf bones), 7 *tarsal* bones (heel), 5 *metatarsals* (sole of foot), and 5 *digits* composed of 14 *phalanges*.

In the pig the bones are similar to those of man. The primary difference is again in the number of digits. The pig has only 4 digits each with 3 phalanges.

HUMAN SKELETON

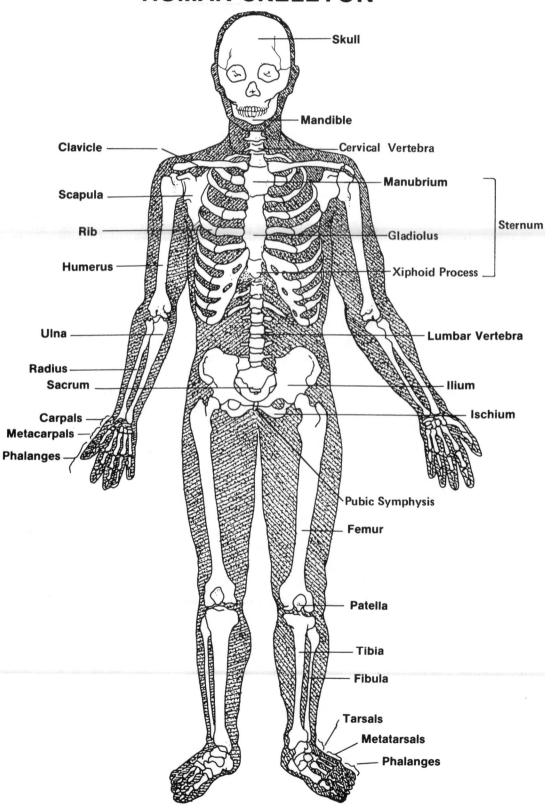

- Skull
- Mandible
- Cervical Vertebra
- Clavicle
- Manubrium
- Scapula
- Sternum
- Rib
- Gladiolus
- Humerus
- Xiphoid Process
- Ulna
- Lumbar Vertebra
- Radius
- Sacrum
- Ilium
- Carpals
- Ischium
- Metacarpals
- Phalanges
- Pubic Symphysis
- Femur
- Patella
- Tibia
- Fibula
- Tarsals
- Metatarsals
- Phalanges

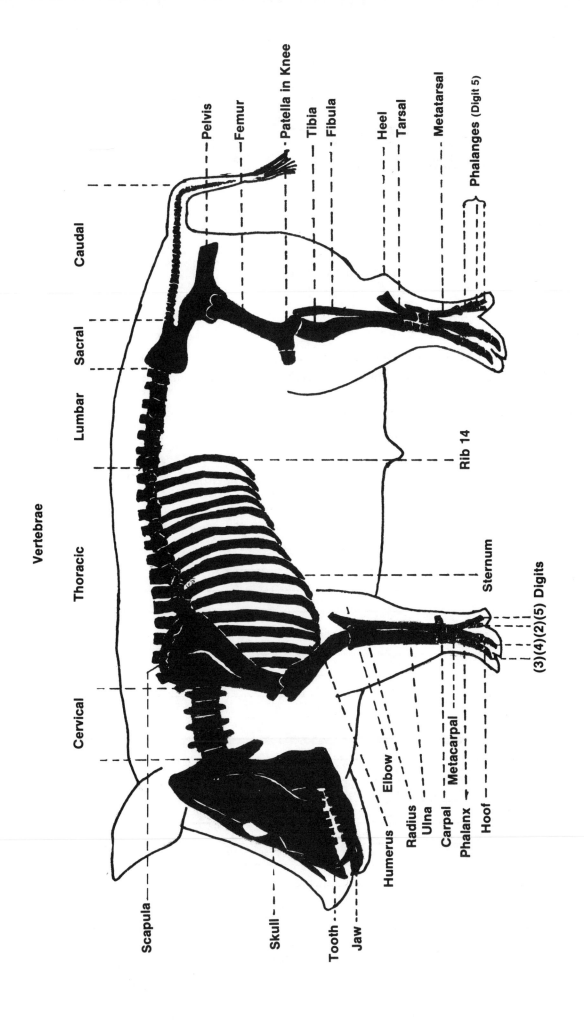

SKELETON OF THE ADULT PIG

15

SELF-QUIZ I
SKELETAL SYSTEM

1. The skull is composed of the cranial and _____ bones.
2. Name the eight cranial bones.
3. Name the main divisions of the axial skeleton.
4. Name the main divisions of the appendicular skeleton.
5. How do the vertebral columns of man and pig differ?
6. Identify the "atlas" and "axis".
7. Name the 3 auditory ossicles.
8. The pelvis is composed of 3 bones: the _____, _____, and _____.
9. a) Which is the only bone of the human body which does not articulate with another bone?
9. b) Name a bone of man missing in the pig.
10. Define each of the following terms listed below:

a) condyle	d) synovium	g) concha	j) lateral malleolus
b) foramen	e) bursa	h) "floating" rib	
c) diarthrosis	f) ethmoid	i) transverse process	

ANSWERS

1. _____
2. _____
3. _____
4. _____
5. _____
6. _____
7. _____
8. _____
9a. _____
9b. _____
10a. _____
10b. _____
10c. _____
10d. _____
10e. _____
10f. _____
10g. _____
10h. _____
10i. _____
10j. _____

Label all of the features indicated in the following diagram.

HUMAN SKELETON

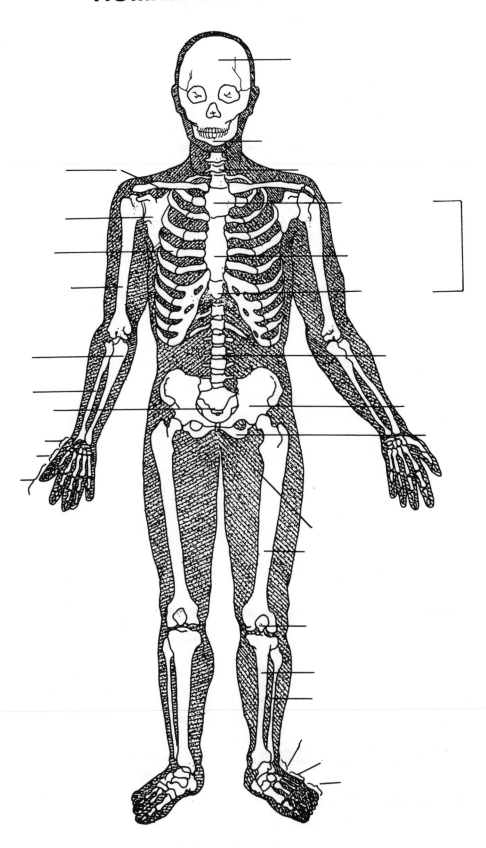

17

GENERAL DISSECTION HINTS

The term "dissection" means more than merely cutting your specimen apart. It is a refined method of seeking, exposing, identifying, and studying the internal anatomy. It helps to bring into view structures not readily seen.

Use your *scalpel* sparingly. Rely primarily on your dissecting needles. They are especially helpful in tracing blood vessels and nerves, and in separating muscles. As organs are exposed, study the associated blood vessels and nerves. Leave them intact unless directed to do otherwise. Check often with the photos in the manual. Confirm the names and location of body parts of your specimen.

When using your *scissors,* advance with the rounded, blunt end, not the sharp pointed end. Your *forceps* should be strong, able to hold on to thick muscle, yet fine enough to grasp narrow nerves. It is advisable to have more than one type of forceps. Move organs aside with your fingers or with a blunt probe.

Greater caution must be exercised in dissecting a fetal animal than an adult animal. The organs and tissues have not yet fully developed. They are therefore, smaller, differently shaped, and more delicate than those of the adult. It will mean using greater care in cutting the soft thin skin, not cutting any of the muscles below the skin, if they are to be studied. A careless movement of finger, scissor, or scalpel may tear, cut, or destroy important structures.

In order to protect yourself against the effects of the preservative solution upon your hands, it is suggested that you apply lanolin or vaseline at the outset or wear thin rubber gloves. Line your dissection pan with paper towels in order to absorb excess fluids, as a storage for structures removed, and to facilitate cleaning up at the close of the session.

At the end of each class, wrap the fetal pig in wet paper towels before returning it to the plastic bag. Twist the top of the bag and close tightly with a rubber band. These procedures will protect your specimen from drying out between dissection sessions. Remove the paper towels lining the dissection pan, together with any structures removed, and discard.

In order to further preserve the softness and texture of pig muscles and organs, apply the following solution with a one inch paint brush at the close of each session:

Carbolic Acid (Phenol) crystals	—	30 grams
Glycerin	—	250 ml.
Water	—	1000 ml.

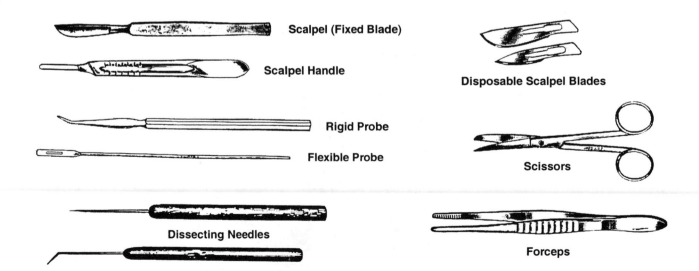

Scalpel (Fixed Blade)

Scalpel Handle

Disposable Scalpel Blades

Rigid Probe

Flexible Probe

Scissors

Dissecting Needles

Forceps

DISSECTION INSTRUMENTS

MUSCULAR SYSTEM - AN OVERVIEW

Skeletal muscles enable the body to move. Most are firmly anchored to the bone at one end, the *origin* of the muscle, while the other end is attached to the bone to be moved, and is known as the *insertion*. The fleshy central portion is termed the *belly*. The ends of a muscle are attached to bone most often by means of a narrow band of connective tissue called a *tendon*. They may also be joined directly to the *periosteum* of the bone. Finally, muscles may be united with each other or to a bone by means of a broad, flat sheet of tendonous tissue known as an *aponeurosis*.

As you dissect, locate the origins and insertions of the muscles studied. Then free the muscle from other muscles and from the nerves and blood vessels associated with it. The fine, transparent connective tissue which binds adjacent muscles is *deep fascia,* while tougher and more fibrous *superficial fascia* connects the skin to the muscles below. When the muscle has been freed, pull it gently. This will duplicate the muscle's normal contraction. Observe which bones or organs are moved, which are relatively stable.

Actions of Muscles:

The *action* of a muscle results from its contraction. Muscles are generally arranged in *antagonistic* pairs. This means that when a muscle causes a structure to move in one direction, one or more antagonists cause it to move in the opposite direction. Some muscles assist others in their actions, thereby bringing about mote efficient movement. These are known as *synergistic* muscles.

Flexion — to bend at a joint decreasing the angle at that joint; examples: elbow or knee joint

Extension — to straighten joint increasing the angle at that joint

Adduction — to move appendage toward sagittal midline; example: lowering arms from shoulder level, to rest at sides

Abduction — to move appendage away from sagittal midline; example: raising arms from rest at sides to shoulder level

Supination — to turn palm of hand upward

Pronation — to turn palm of hand downward

Rotation — to move a structure about a point; example: turning head from side to side

Circumduction — when the distal end of a limb describes a circle while the proximal end remains fixed, as the vertex of a cone; example: the movement of the extended arm in drawing a circle on the blackboard

BEGINNING THE DISSECTION

At this point, some instructors may choose to begin the examination of the oral, the abdominal, and the thoracic cavities. Thus, in many introductory courses, the study of the muscles is skipped entirely. Others may choose to proceed first to a study of the musculature.

In this manual both approaches are satisfied. We shall begin with a study of muscles and follow this with an examination of the oral, abdominal, and thoracic cavities.

Skinning the Pig

Tie the animal to the dissection pan with limbs extended, ventral surface upward. Use your scalpel, forceps, and blunt probe. Proceed as in the accompanying photo, p. 22.

1. Make a mid-ventral incision in the skin from the jaw to the umbilical cord. Be careful to cut the skin only, not the underlying delicate muscle tissue. Do *not* remove the skin from the head (face or skull).
2. Cut around the umbilical cord and proceed posteriorly as in the photo to a point just anterior to the external gentalia.
3. As you cut with the scalpel, lift the skin with your forceps and separate the skin from the underlying muscles. You will note that the two are held together by a white fibrous connective tissue known as the *superficial fascia*. Cut the fascia as you loosen the skin.
4. Continue to reflect the skin toward the dorsal side. Use the back of your scalpel, a blunt probe, or your fingers to facilitate separation.
5. Cut along the medial surface of the forelimbs and hind limbs and extend to the wrists and ankles.
6. Leave the skin intact around the urogenital and anal areas (*perineum*). Remove the skin from the proximal 1/3 of the tail.
7. Turn the pig over. Complete the skinning of the limbs and the entire dorsal surface from the base of the skull, the neck, dorsal thorax and abdomen, to the proximal 1/3 of the tail. Do not discard the skin. Use it to wrap the pig, in addition to wet paper towels at the close of each dissection session. The only areas still covered by skin are the head, feet, perineum, and the distal portion of the tail.

Your pig should now appear as the one in the photo, p. 26.

Look for light brown fibers adhering tightly to the underside of the skin. These are *cutaneous muscles*. They include the:

Cutaneous Maximus — This muscle covers most of the sides of the body in the thoracic and abdominal areas. It serves to twitch the skin to avoid irritants. It originates from muscles in the axilla, the thorax, and abdomen, and inserts on the skin. It is not found in man.

Platysma — This is another cutaneous muscle. It is found on the lateral surface of the head and neck. It moves the skin on the neck and face. It originates from the mid-dorsal area over the neck and inserts in the skin of the face near the ears, eyes, and mouth.

As you continue the dissection trim the fascia, other connective tissue and fat covering the muscles. Observe the direction in which the muscle fibers lie. The fibers of a single muscle are generally oriented in only one direction. Look for natural separations between muscles, then slit the fascia between muscles with a dissecting needle or a scalpel. Caution: Do not cut the muscle fibers, only the connective tissue between them.

When the study of superficial muscles has been completed, *transect* the muscles. This is done by cutting them at right angles to the directions of the fibers at the belly area and folding them back, or *reflecting* them, to their origins and insertions. This will reveal the deeper muscle layers.

We shall begin by examining the superficial muscles first on the ventral, then the lateral and finally the dorsal surface. Some deep muscles will also be studied. A Self-Quiz for students will be found at the end of the unit on the Muscular System.

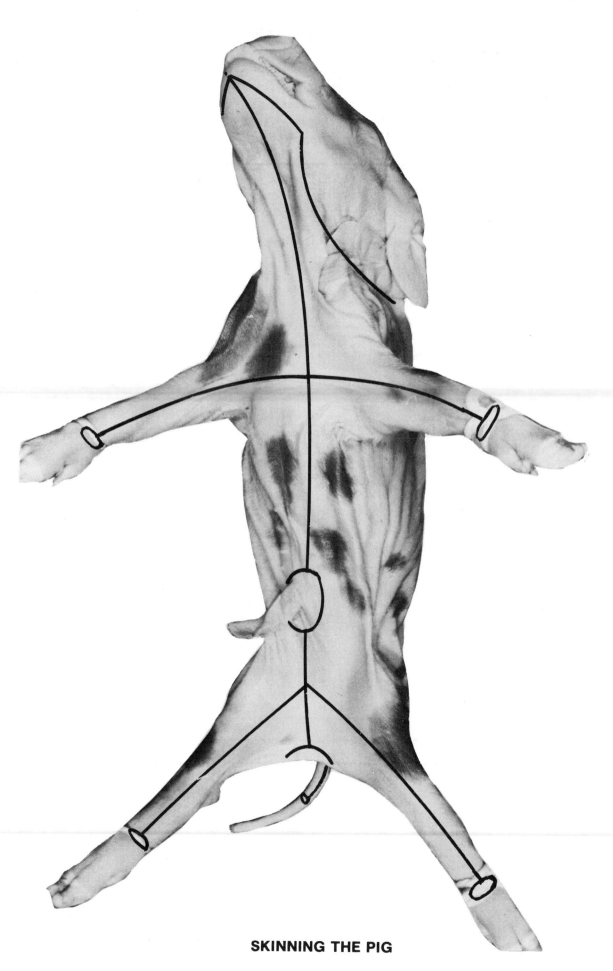

SKINNING THE PIG

SUPERFICIAL MUSCLES - VENTRAL VIEW

Turn the pig to expose the ventral surface as in the accompanying photo. We shall first study the muscles of the ventral thorax and abdomen. See the closeup photo of this area, p. 27.

Thoracic Muscles

The chest area is dominated by the *pectoral* muscles. As in man, there are two of these muscles. Their relative sizes and positions are different from those in man.

Pectoralis Major *(Pectoralis Superficialis)* — This is a broad fan-shaped muscle. It originates from the upper portion of the sternum and inserts along almost the entire length of the humerus and on the fascia covering the proximal end of the forearm. It acts to draw the forelimb towards the chest (adduction). The *cephalic vein* may be seen at the distal end of this muscle extending from the shoulder over the forearm. This major vein drains the superficial musculature of the forelimb.

Pectoralis Minor *(Pectoralis Profundus)-(Posterior Deep Pectoral)*—As the names in parentheses indicate, this muscle is posterior and lies deep to the pectoralis major.

In humans, however, it is the smaller of the two. It arises from the entire length of the sternum. Its fibers extend anteriolaterally deep to those of the pectoralis major. Insertion: its posterior fibers insert on the proximal end of the humerus, its anterior fibers extend on the shoulder joint. Action: To adduct and retract the forelimb.

Latissimus Dorsi — A broad muscle with an extensive origin on the dorsal surface. It is very prominent on the lateral and ventral surfaces too. Some of its fibers originate in the lumbodorsal fascia, others from the lumbar and the last four thoracic vertebrae. The muscle is directed anteriorly, lies on the lateral surface of the thorax, becomes narrower, comes lie ventral, and inserts on the proximal end of the humerus on its medial serface. Action: Moves the forelimb dorsally and posteriorly. Because of its extensive origin and narrow insertion, it gives the forelimb great power.

Abdominal Muscles

The abdominal area is not protected by a bony structure as is the thorax. The abdominal organs are held in place by the pressure of the lateral abdominal muscles. They compress the abdominal wall and aid flexion of the trunk. The abdominal muscles include the:

External Oblique — This is a thin broad sheet of muscle covering the ventral and lateral abdominal surfaces. It is the outermost of the three lateral abdominal layers. It originates on the posterior ribs and the *lumbodorsal fascia*, an aponeurosis on the dorsal surface, and inserts on an aponeurosis along the mid-ventral surface. The *linea alba*, a white line of connective tissue along the mid-ventral surface represents the fusion of the aponeuroses of the right and left sides. The fibers of the external oblique extend caudally and ventrally in an oblique direction across the abdominal surface.

Internal Oblique — Lift the edge of the external oblique where it joins the aponeurosis as in the photo, and expose the second layer of abdominal muscles, the internal oblique. Its fibers run in a direction opposite to those of the upper layer, namely, ventrally and anteriorly.

Transversus Abdominis — This is the innermost of the abdominal muscle layers. Its fibers extend ventrally and slightly caudally, almost parallel to those of the external oblique. It arises from the lower rib and the lumbar vertebrae and inserts along the linea alba by an aponeurosis. The arrangement of the fibers of the three layers gives the abdominal wall its strength. Below the transversus abdominis lies the thin glistening membrane, the *parietal peritoneum* which lines the abdominal cavity.

The three layers of abdominal muscles are separated and clearly visible in the closeup photo, page 27.

Rectus Abdominis — In the mid-ventral area, on either side of the linea alba, lie two parallel muscles. They extend from the pubis cranially to insert on the upper ribs and sternum. For much of their course they lie between the aponeurosis of the internal oblique and the transversus abdominis.

Latissimus Dorsi — Although this is primarily a muscle of the thoracic region, it is very prominent on the abdominal ventral surface as well. It arises from aponeuroses along the mid-dorsal line of the posterior thoracic region and from most of the lumbar region. It covers the lateral surface of the body in this area. It extends ventrally to insert on the humerus. It gives to the humerus great power for pulling backward when the pig is running.

Thigh

Two broad, thin superficial muscles are visible in the anterioventral thigh area. The *femoral artery* can be seen passing between them. They are the:

Sartorius — This muscle occupies the anterior half of the thigh. It resembles a flattened band about ½ inch wide. Separate it from the neighboring muscles. Origin: the iliac. Insertion: the proximal end of the tibia. Action: adducts the thigh and extends the lower hindleg.

Gracilis — The second medial superficial thigh muscle is also broad and thin. It covers the posterior position of the medial (inner) thigh. Its origin is near the pubic symphasis of the pubis and inserts upon the proximal third of the tibia on the medial side.

Shoulder and Neck Muscles

Brachiocephalic — This large prominent muscle seen in the photos, pages 26 and 28, extends from the back of the neck, the mastoid process and the back of the head. It lies upon the shoulder extending to the humerus. Its action is to move the forelimb anteriorly.

In those mammals where a clavicle is present, this muscle is divided into two. The anterior portion extends from the neck to the clavicle and is known as the *clavotrapezius*, while the posterior portion extending from the clavicle to the humerus is known as the *clavodeltoid*, or *clavobrachialis*. The pig, however, has no clavicle, thus the brachiocephalic is a single muscle.

In the mid-ventral neck area, several muscles may be seen. Use your needle probe to separate some of these narrow bands of muscle.

Sternomastoid (Sternocephalic) — This is a large "V"-shaped band of muscle on the ventral and lateral surfaces of the neck. It extends from the anterior position of the sternum to the mastoid process. It acts to flex the head upon the chest and to incline it to one side.

Two other bands of muscles in this area are the *cleidomastoid* and the *omohyoid*.

Several narrow bands of muscle cover the throat. They extend from the sternum, to the hyoid and larynx on the ventral side of the neck. The names of these muscles indicate their origins and insertions. They include the:

Sternohyoid — This is the most ventral of these muscles. It runs along the mid-ventral line in an anterior to posterior direction.

Sternothyroid — Bisect and reflect the sternohyoid and the sternothyroid will be seen deep and lateral to it. It covers the thyroid gland and trachea.

Thyrohyoid — This is a shorter muscle which extends from the insertion of the sternothyroid to the hyoid.

Digastric — Below the inner edge of the mandible one can see the "V"-shaped digastric muscle. It extends from the occipital and temporal bones to the mandible and acts to lower the jaw.

Mylohyoid — The fibers of this muscle run transversely between the digastric muscles in the mid-ventral area. The ends of the muscle pass deep to the digastric. It acts to raise the floor of the mouth.

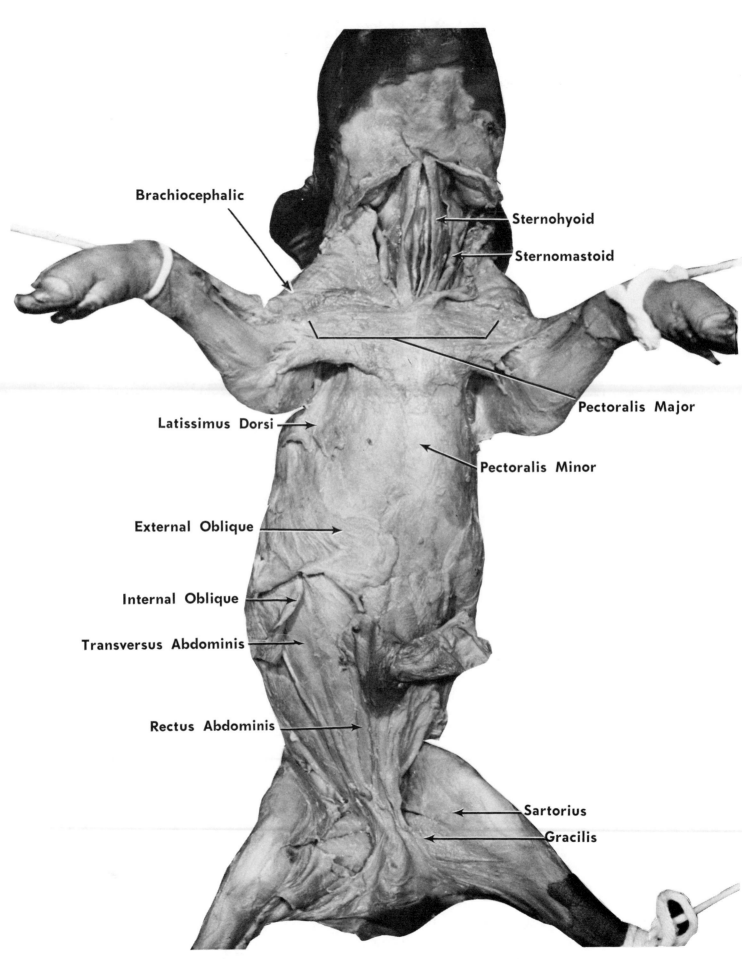

Brachiocephalic

Sternohyoid

Sternomastoid

Pectoralis Major

Latissimus Dorsi

Pectoralis Minor

External Oblique

Internal Oblique

Transversus Abdominis

Rectus Abdominis

Sartorius

Gracilis

SUPERFICIAL MUSCLES, Ventral View

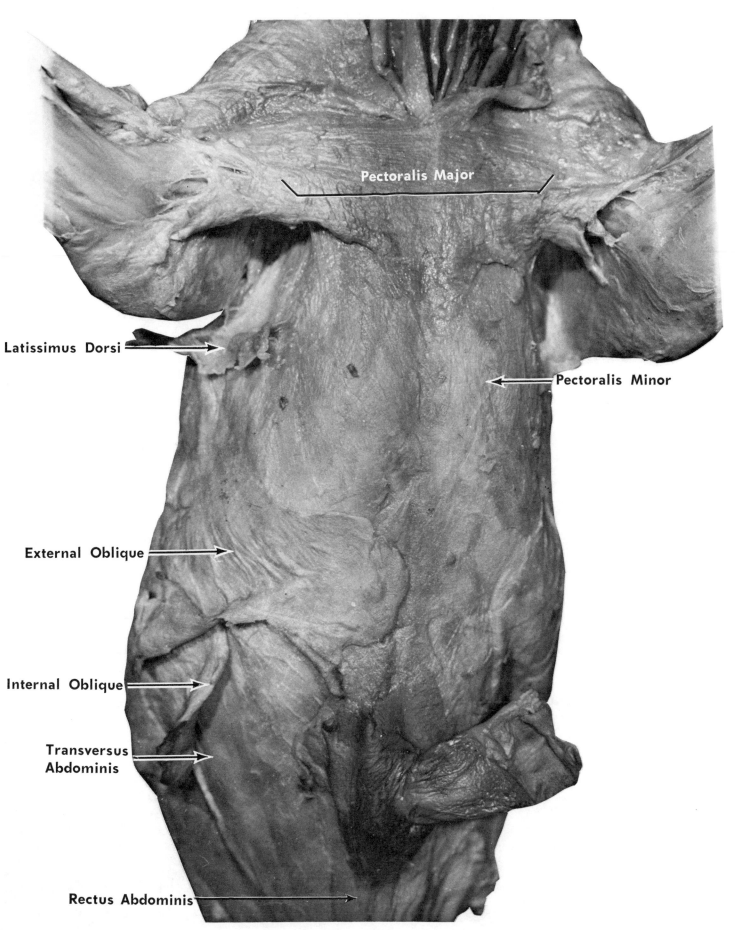

Pectoralis Major

Latissimus Dorsi

Pectoralis Minor

External Oblique

Internal Oblique

Transversus
Abdominis

Rectus Abdominis

THORAX AND ABDOMEN, VENTRAL VIEW (Close-up)

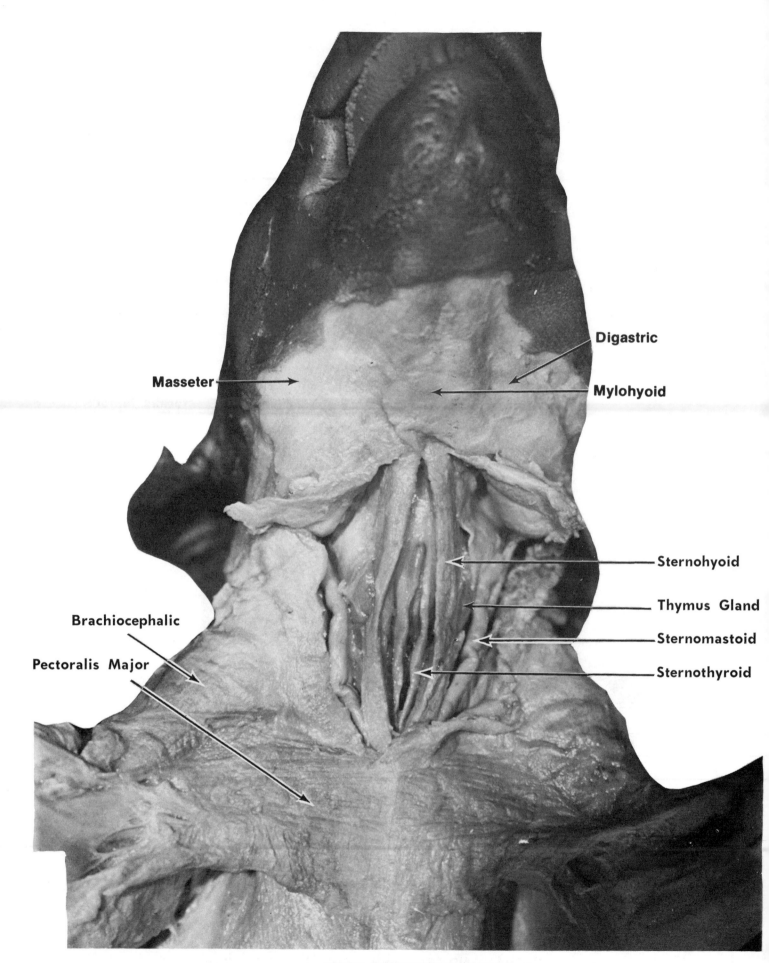

Digastric

Masseter

Mylohyoid

Sternohyoid

Thymus Gland

Brachiocephalic

Sternomastoid

Pectoralis Major

Sternothyroid

NECK, VENTRAL VIEW

28

SUPERFICIAL MUSCLES - LATERAL VIEW

Turn the pig so that it lies on its side as in the photo on page 32.

Shoulder Area:
Examine the shoulder area. Separate some of the muscles here by finding the natural separations between them. The following muscles are seen in close-up in the photo, p. 33.

Deltoid Group — This group of muscles is located dorsal and anterior to the base of the forelimb. It is composed of three muscles in the pig and one in man.

Brochiocephalic — The posterior portion of this muscle covers the ventral shoulder area. This large muscle has been described earlier. See p. 24.

Acromiodeltoid — An elongated slender muscle extending over the anterior surface of the shoulder to the forelimb. Origin: scapula. Insertion: into proximal end of humerus. Action: to raise the humerus.

Spinodeltoid — This muscle is larger than the preceding one and lies posterior to it. It originates from the spine of the scapula and inserts into the proximal end of the humerus. It acts in unison with the acromiodeltoid.

The Rib Cage Area:
Intercostal — These muscles are located between the ribs. They are seen in the close-up photo on p. 47. They are major muscles of *respiration*. The intercostals serve to raise and lower the ribs and thereby to expand and contract the chest cavity. They are composed of two sets.

External Intercostals — Their fibers are directed in a caudo-ventral direction. As their name indicates, they are the more superficial of the two. These muscles raise the rib cage during inhalation. They cover the dorsal portion of the rib cage but are absent ventrally.

Internal Intercostals — Their fibers are directed in a caudo-ventral direction, at right angles to those of the external intercostals. They are the deeper rib muscles. They lower the rib cage during expiration. The internal intercostals are located along the entire interspace between the ribs, from the dorsal side to the mid-ventral sternum.

The Forelimb:
Near the shoulder we can identify a large muscle, the triceps brachii, which covers almost the entire superficial surface of the forelimb. It is divided into two muscle masses upon the humerus. These are the:

Triceps Brachii, Long Head — This is the largest segment of the triceps group. It is a triangular shaped muscle on the dorso-lateral aspect of the forelimb. It arises from the posterior border of the scapula.

Triceps Brachii, Lateral Head — This flattened segment lies upon the dorsal portion of the forelimb. It originates from the proximal end of the humerus in its lateral surface.

Triceps Brachii, Medial Head — This segment is deep to the other two and can not be seen superficially. It also arises from the proximal portion of the humerus, from the medial side.

All three heads of the triceps brachii insert in common upon the olecranon process of the ulna, and act together to extend the forelimb at the elbow.

The Lower Foreleg:

On the lateral surface, a grop of thin muscle bands may be separated on the lower foreleg. These are extensors of the foot and digits. They include the following muscles:

Extensor Carpi Radialis
Extensor Digitorum Communis
Extensor Carpi Ulnaris
Extensor Digitorum Lateralis

The origin of these muscles is in the distal portion of the humerus and the proximal portion of the radius and ulna. Their insertions are upon the digits or the metacarpal bones.

These muscles are best seen in the close-up view, p. 33.

Hip and Thigh:

Examine the lateral muscles of the hip and thigh. These can best be seen in the photo, p. 34.

Tensor Fasciae Latae — This muscle arises from the crest of the ilium. As its name indicates it "tenses" or pulls upon the *fascia lata*, a white sheet of fascia near the knee and over the tibia which serves as its insertion. Thus, in addition to tensing the fascia lata, the muscle also acts to flex the hip and extend the knee.

Gluteus Maximus — (Gluteus Superficialis) — In the pig, this is a relatively small muscle, while in humans it forms the primary tissue of the buttocks. Is is a thin muscle which lies upon the hip posterior to the tensor fascia latae. It originates from the last sacral and first caudal vertebrae and inserts into the fascia lata. Action: abductor of the thigh.

Gluteus Medius — This muscle is thicker and narrower than the gluteus maximus. It is also more conspicuous. It lies deep to the gluteus maximus. It originates from the lumbodorsal and gluteal fascia and inserts upon the greater trochanter of the femur. It acts to extend the hip and abduct the hind limb.

The Hamstrings — The following three muscles are collectively known as the hamstrings. The name originates from the practice of butchers who hung hams by the tendons of these muscles. They include the: *Biceps Femoris, Semi-tendinosus,* and *Semi-membranosus.*

Biceps Femoris — This very broad, thick muscle covers most of the lateral surface of the thigh. It lies posterior to the tensor fascia latae. It originates from the posterior portions of the sacrum and ischium and inserts by an aponeurosis along the tibia. It retracts the knee, flexes the shank, and abducts the thigh.

Semitendinosus — Although portions of this muscle and the following one may be seen superficially, their major portion is hidden. Bisect the biceps femoris and reflect its ends. Note the *obturator* and *posterior* femoral arteries and the very thick *sciatic nerve*, which innervates many of the hip and leg muscles. Origin: the anterior caudal vertebrae and ilium. Insertion: Proximal end of tibia, the fascia of the leg, and the calcaneous.

Semimembranosus — This is another large muscle of the thigh. It lies posterior and medial to the semitendinosus. It originates on the iscial tuberosis and inserts upon the distal end of the femur and the proximal end of the tibia. It acts to extend the hip and adduct the hind limb. You will again see this muscle when you are dissecting the medial surface of the thigh.

Lower Hindleg:

Gastrocnemius — This is the large muscle of the calf. It originates as two separate heads, the *lateral head* and *medial head*, upon the distal end of the femur. It inserts upon the heel bone, the *calcaneous*, by way of the long, tough, Achilles tendon. It acts as an extensor of the ankle.

Soleus — This muscle lies deep to the gastrocnemius. It originates upon the fibula and inserts together with the gastrocnemius upon the calcaneous by way of the Achilles tendon. Thus, the Achilles tendon serves both muscles. Action: to extend the ankle or to flex the knee.

Tibialis Anterior — This is the most ventral muscle of the shank. As its name indicates it lies upon the tibia. It originates from the proximal end of the tibia and fibula and inserts upon the second metatarsal.

Peroneus: Longus, Brevis, and Tertius — This group of deeper muscles of the lower hind leg act to flex the ankle. They originate from the tibia and the distal end of the femur, and insert on the metatarsal and tarsal bones.

Extensor Digitorum Longus, Flexor Digitorum Longus — These are two of several extensor and flexor muscles found in this area. They insert upon and move the metatarsals and digits. They are shaped like narrow bands and end as long, tough tendons, some branching to several digits.

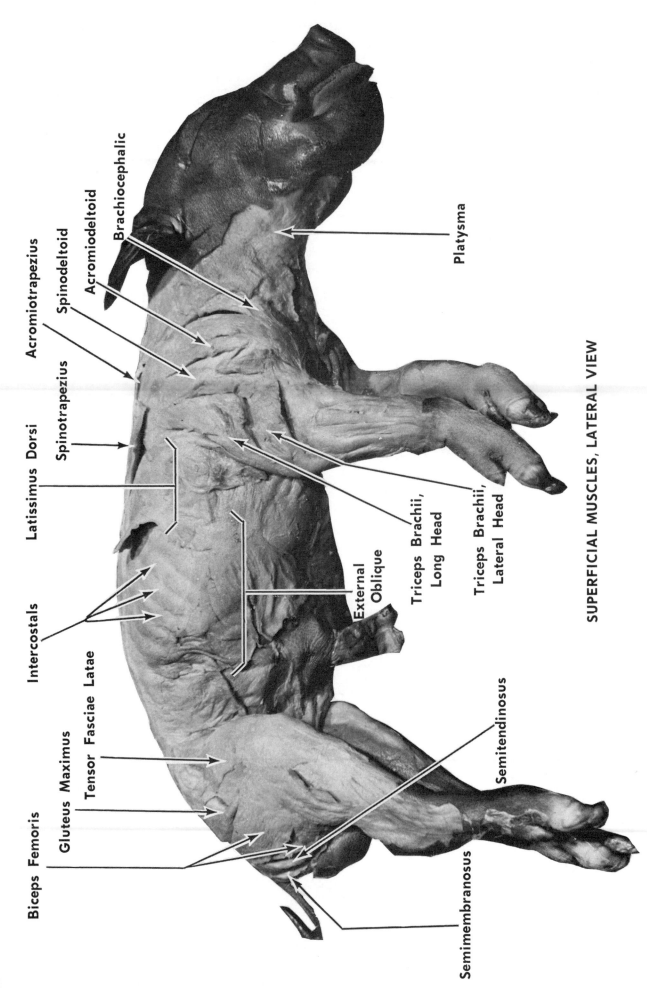

Brachiocephalic

Acromiodeltoid

Spinodeltoid

Acromiotrapezius

Spinotrapezius

Latissimus Dorsi

Intercostals

Biceps Femoris

Gluteus Maximus

Tensor Fasciae Latae

Platysma

Triceps Brachii, Long Head

Triceps Brachii, Lateral Head

External Oblique

Semitendinosus

Semimembranosus

SUPERFICIAL MUSCLES, LATERAL VIEW

32

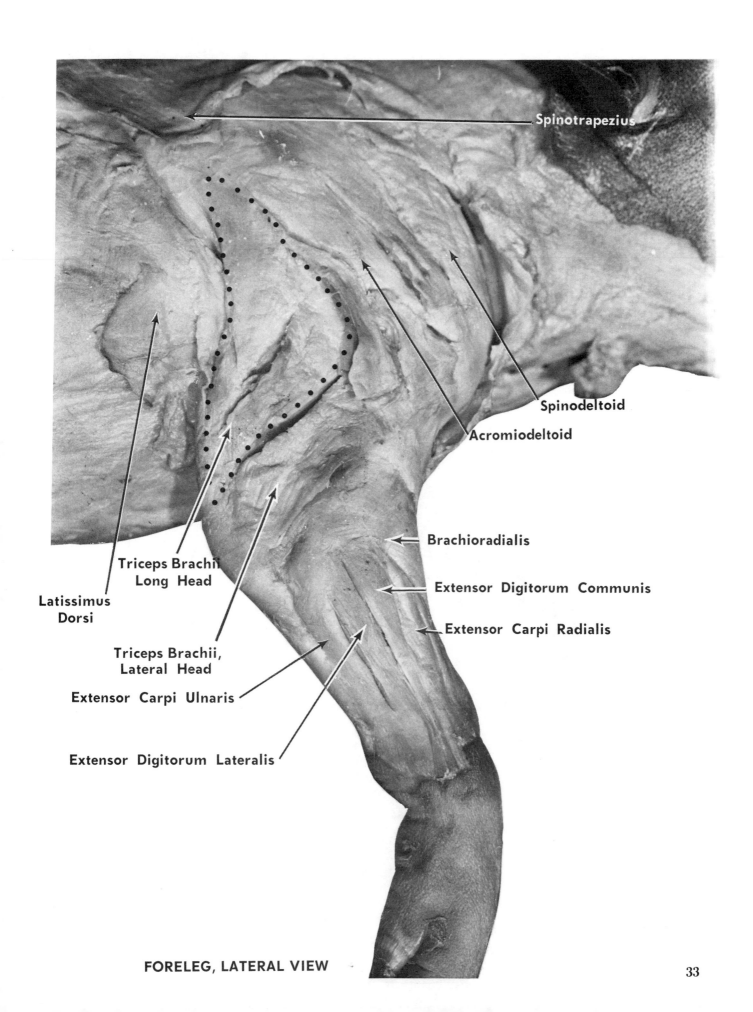

Spinotrapezius

Spinodeltoid

Acromiodeltoid

Brachioradialis

Extensor Digitorum Communis

Extensor Carpi Radialis

Triceps Brachii
Long Head

Latissimus
Dorsi

Triceps Brachii,
Lateral Head

Extensor Carpi Ulnaris

Extensor Digitorum Lateralis

FORELEG, LATERAL VIEW

33

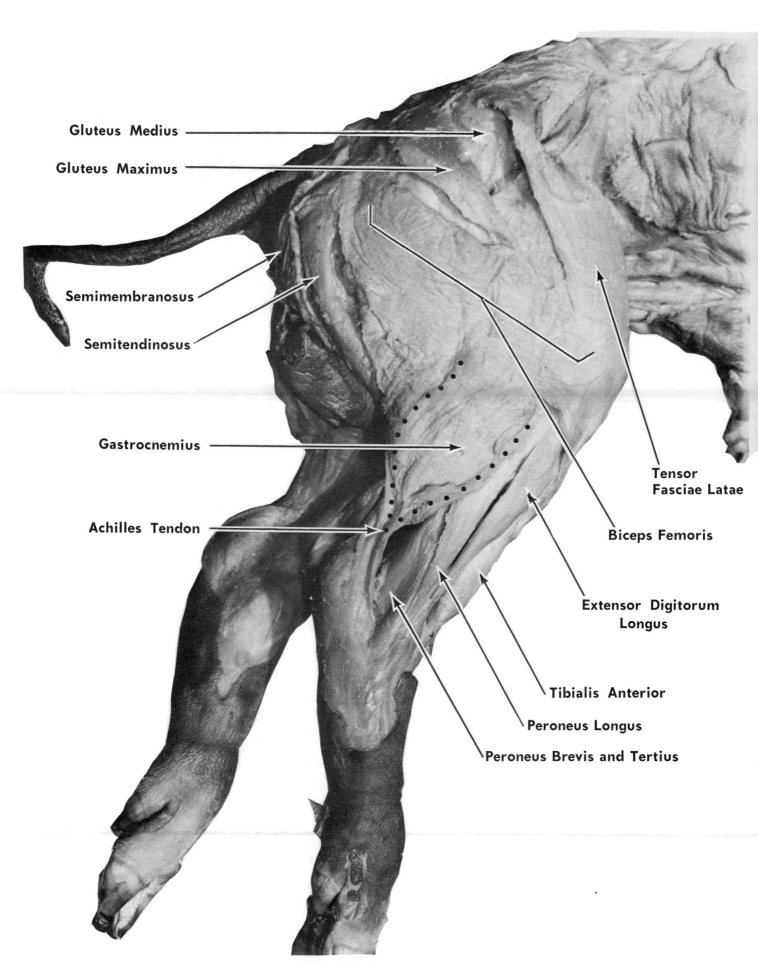

Gluteus Medius

Gluteus Maximus

Semimembranosus

Semitendinosus

Gastrocnemius

Achilles Tendon

Tensor
Fasciae Latae

Biceps Femoris

Extensor Digitorum
Longus

Tibialis Anterior

Peroneus Longus

Peroneus Brevis and Tertius

HIND LEG, LATERAL VIEW

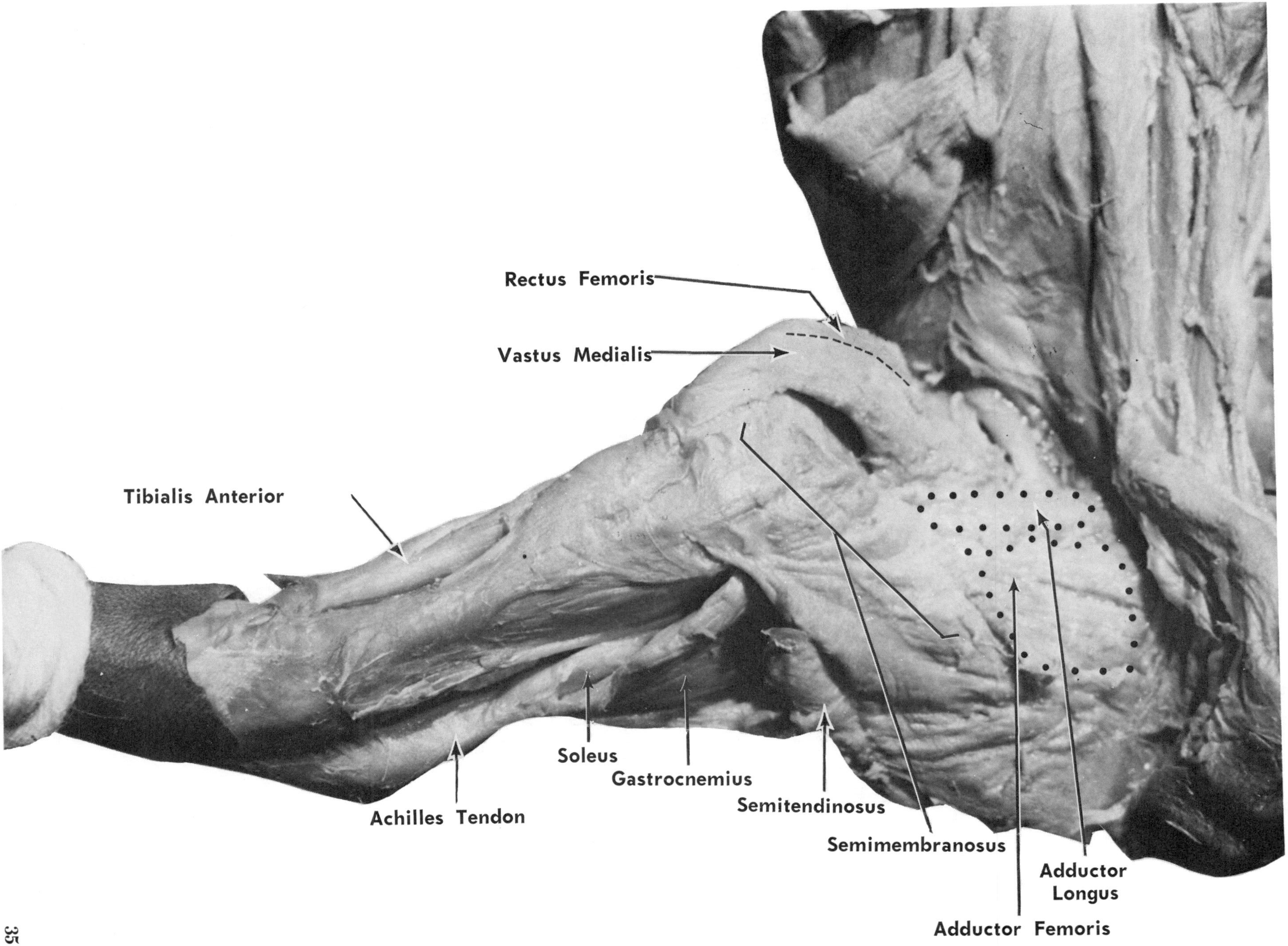

HIND LEG, MEDIAL VIEW

Adductor Longus

Adductor Femoris

Semimembranosus

Semitendinosus

Gastrocnemius

Soleus

Achilles Tendon

Rectus Femoris

Vastus Medialis

Tibialis Anterior

SUPERFICIAL MUSCLES - DORSAL VIEW

Turn the pig over to expose the dorsal surface as in the photo.

Neck and Shoulder:

Trapezius — This is an extensive superficial muscle. In man, this is a simple broad dorsal superficial muscle covering the shoulder and anterior thorax. In the pig it is divided into three separate muscles. They are:

Brachiocephalic (clavotrapezius) — This is the most anterior of the three. As described earlier, (see p. 24) the proximal portion of the brachiocephalic covers the dorsal surface of the neck. It originates from the back of the skull and extends to the humerus. It acts to move the forelimb anteriorly.

Acromiotrapezius — This second muscle of the trapezius group is located posterior to the brachiocephalic. It is a thin, broad, fan-shaped muscle originating on the cervical vertebrae. It inserts upon the spine of the scapula by means of a broad aponeurosis. Action: to pull the scapula to the mid-dorsal line.

Spinotrapezius — This is the most posterior of the three. It covers most of the dorsal thorax. It corresponds most closely to the location and shape of the trapezius in man. It originates from the spines of the first ten thoracic vertebrae and is inserted into the scapula. It also pulls the scapula mid-dorsally.

Latissimus Dorsi — This muscle, located posterior to the spinotrapezius, although originating and covering an extensive portion of the dorsal surface, extends to the lateral and ventral sides to insert upon the humerus. It has been described earlier (see p. 23) when the pectral muscles of the ventral surface were discussed.

An extensive aponeurosis, *the lumbodorsal fascia*, covers the lower back to join the superficial muscles in that area.

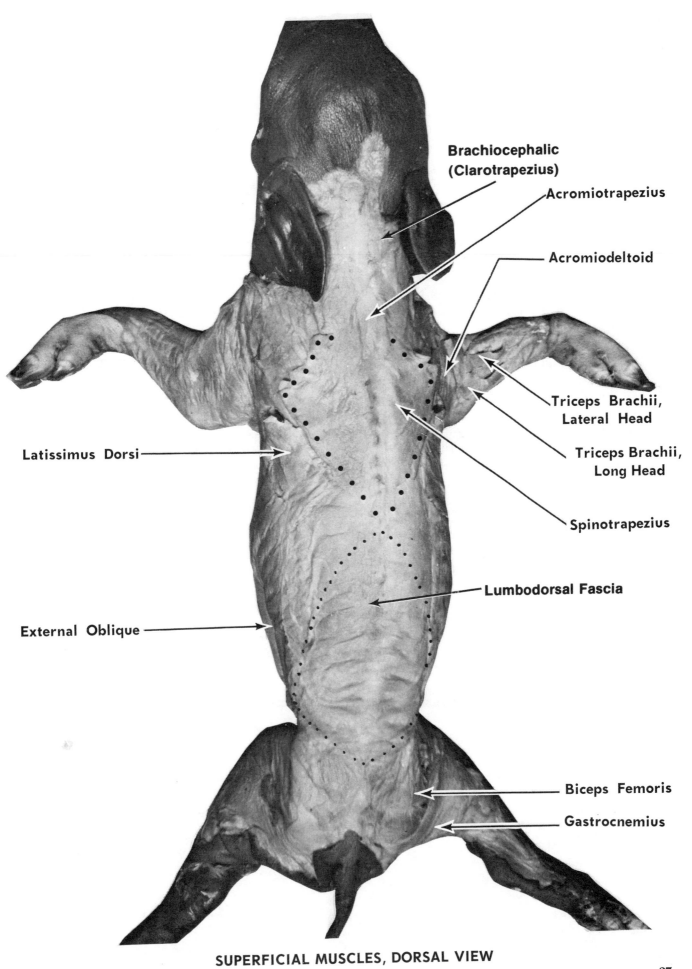

Brachiocephalic (Clarotrapezius)

Acromiotrapezius

Acromiodeltoid

Triceps Brachii, Lateral Head

Triceps Brachii, Long Head

Latissimus Dorsi

Spinotrapezius

Lumbodorsal Fascia

External Oblique

Biceps Femoris

Gastrocnemius

SUPERFICIAL MUSCLES, DORSAL VIEW

DEEP MUSCLES - INTRODUCTION

The dissection of the deeper muscles should be done on one side only, leaving the superficial muscles on the second side intact. Compare the superficial muscles on one side the with deeper muscles on the other side. This procedure permits later utilization of superficial muscles for review and study. Also, if for some reason the deeper muscles on one side are improperly dissected or destroyed, those of the second side are then still available. Similarly, dissect the deeper muscles of only one forelimb and one hind limb, leaving the superficial muscles intact upon the second limb.

In order to expose the deeper muscles it is best to *transect* each of the superficial muscles. This is done by cutting them at right angles to the direction of the fibers at the central belly area. They are then folded back, or *reflected*, to their origins and insertions. They can thus easily be "reconstructed" at any time in order to examine the relationship of the deeper muscles to those of the superficial layers.

You will find that the deeper muscles are quite different from those above them. The variations in shapes, lengths, directions of the fibers, as well as their actions are revealed only after they have been exposed. It is impossible to guess about them by viewing the superficial muscles alone. For example, could we predict the unusual shape of the serratus ventralis or dorsalis muscles by merely viewing the superficial thoracic musculature? Such examples abound when considering the entire pig musculature.

The description of the deeper muscles will repeat the pattern followed for the superficial muscles; first the anterior ventral area, then the posterior, finally the deeper muscles of the dorsal region. A Self-Quiz for students is found at the end of the entire unit on muscles.

DEEP MUSCLES - VENTRAL VIEW

Forelimb:

Bisect the pectoralis major and minor muscles and reflect their ends. Also, cut the latissimus dorsi near its insertion.

This will expose the deeper muscles, blood vessels, and nerves of the forearm as in the photo p. 41.

Biceps Brachii — Although in man this muscle is superficial, and prominent, in the pig it lies deep to the pectoral and brachial muscles. It is the primary flexor of the antebrachium (lower foreleg). It lies upon the anterio-medial surface of the humerus. In man the muscle has two heads (biceps) while in the pig it arises by means of a single tendon that passes over the humerus to insert upon the coracoid process of the scapula. It inserts upon the radius and ulna.

Brachialis — This muscle lies lateral to the biceps brachii. The insertions of the brachiocephalic and pectoral muscles pass between them. It originates from the radius and inserts upon the ulna. They are both flexors of the lower forearm.

Triceps Brachii, Medial Head — Although the *long head* and the *lateral head* of the *biceps brachii* muscle have already been described earlier (see p. 29) in discussing the superficial muscles, the *medial head* of this muscle could not be seen superficially. It may be viewed here, as in the photo, p. 41.

In the lower forelimb several flexor and extensor muscles of the carpals, metacarpals, and digits are seen. The flexors are generally found on the medial side, the extensors on the lateral side. The extensors on the lateral surface were already described on page 30, and seen in the photo, p. 33. The flexors include the:

> **Flexor Carpi Brachialis**
> **Flexor Digitorum Profundus**
> **Flexor Carpi Ulnaris**
> **Palmaris Longus**

Use your dissecting needle to separate these narrow muscle bands. They can be seen in the photo on page 41.

The origin of the flexor muscles is in the distal portion of the humerus and proximal portion of the radius and ulna. Their insertions are upon the carpals, metacarpals or digits. Trace some of the tendons to their insertions.

Thigh:

Expose the medial surface of the thigh as seen in the photos, pages 35 and 42.

Separate the superficial *sartorius* and *gracilis* muscles. Bisect them and reflect their ends. The following muscles may be seen.

Quadriceps Femoris — This is a group of four anterior thigh muscles. They join to form a common tendon which passes over the patella (knee cap) to insert on the tibia. The portion of the tendon between the patella and the tibia is known as the *patellar ligament*. They act together to extend the hind limb. This muscle group is similar to the one in man.

Rectus Femoris — The first of this group is a thick muscle on the anterior medial side of the thigh. It originates from the ilium.

Vastus Medialis — A small mass of muscle adhering closely to the rectus femoris. It originates from the head of the femur.

Vastus Lateralis — This muscle is best seen on the lateral surface after the *biceps femoris* and *tensor fascia latae* muscles have been cut and reflected. It originates on the lateral surface of the femur.

Vastus Intermedius — This is the last of the quadriceps group. It lies deep to the rectus femoris, between the vastus lateralis and vastus medialis. It is not seen in the photograph. It originates from the ventral surface of the femur.

Once the sartorius and gracilis have been reflected, other muscles may be seen. They include the:

Adductor Magnus — As the name implies, this triangular muscle is an adductor of the thigh. It lies anterior to the semimembranosus and originates from the ventral surface of the pubis and ischium. It is not subdivided into *adductor longus* and *adductor femoris* portions as in humans and other mammals. Its insertion is along much of the length of the femur on its medial side.

Pectineus — This is a smaller triangular shaped muscle, lying adjacent and anterior to the adductor magnus. It originates from the ventral portion of the pubis and inserts upon the femur. It too is an adductor of the thigh.

> **Semimembranosus**
> **Semitendinosus**
> **Gastrocnemius**
> **Soleus**
> **Tibialis Anterior**

The muscles named above are found on the medial surface of the hind leg and may be seen in the photos on pages 35 and 43. However, they have already been described in discussing the lateral muscles of the hind leg (pages 30 to 31).

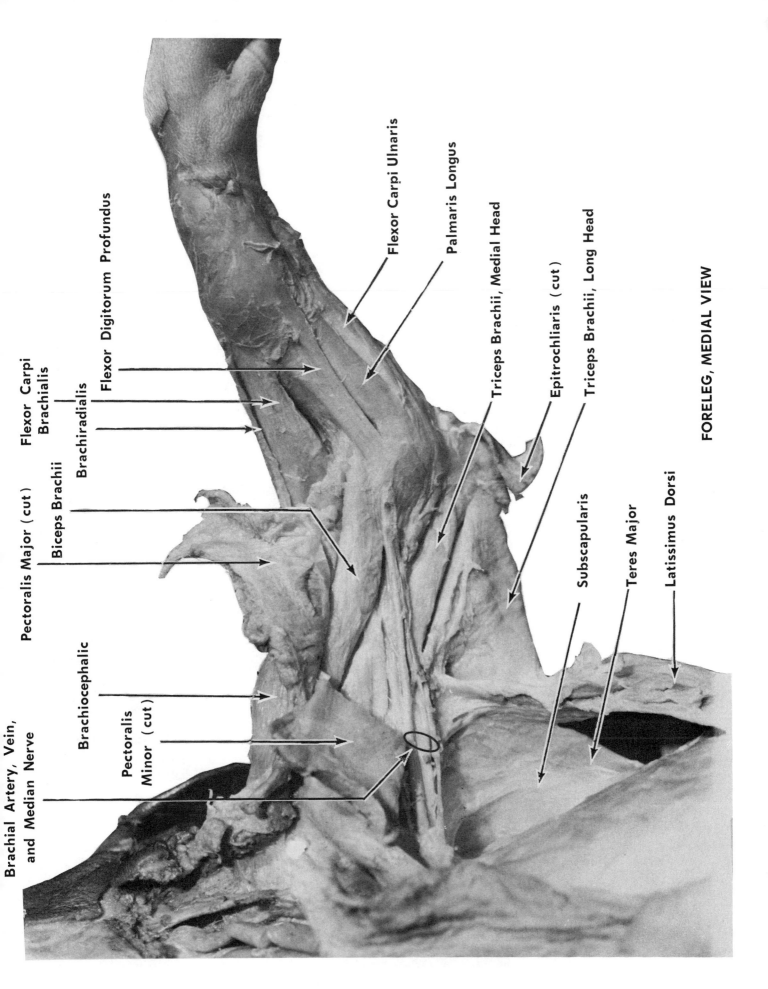

Brachial Artery, Vein,
and Median Nerve

Pectoralis Major (cut)

Flexor Carpi
Brachialis

Brachiradialis

Flexor Digitorum Profundus

Biceps Brachii

Brachiocephalic

Pectoralis
Minor (cut)

Flexor Carpi Ulnaris

Palmaris Longus

Triceps Brachii, Medial Head

Epitrochliaris (cut)

Triceps Brachii, Long Head

Subscapularis

Teres Major

Latissimus Dorsi

FORELEG, MEDIAL VIEW

41

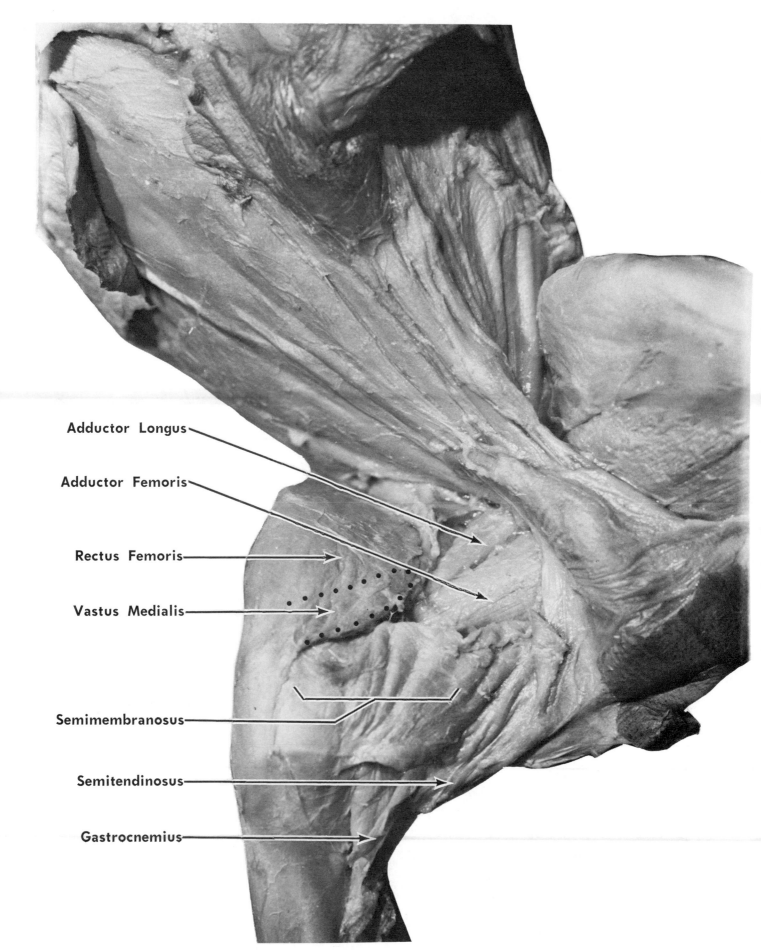

Adductor Longus

Adductor Femoris

Rectus Femoris

Vastus Medialis

Semimembranosus

Semitendinosus

Gastrocnemius

THIGH, MEDIAL VIEW

DEEP MUSCLES - SHOULDER AND THORAX

Ventral View

When the pectoral muscles have been cut and reflected, the scapula can be observed.

Subscapularis — This thick muscle covers almost the entire ventral surface of the scapula. It inserts upon the proximal end of the humerus and acts as an adductor of the forelimb.

Teres Major — This muscle originates upon and covers the axillary posterior borders of the scapula. It inserts upon the humerus by means of a tendon in common with the latissimus dorsi. Its action is to rotate and flex the humerus.

Teres Minor — This muscle is much smaller than the teres major. It originates on the axillary border of the scapula and inserts on the greater tuberosity of the humerus. It assists the infraspinatus muscle in rotating the humerus laterally. It also flexes the shoulder and adducts the forelimb. It is best seen in dorsal view.

Serratus Ventralis — This is a large fan-shaped muscle. Its name is derived from the saw tooth-like edges of the muscle strips. It arises by a number of slips from the lateral surface of the upper eight or nine ribs and from the transverse processes of the last five cervical vertebrae to insert upon the dorsal margin of the scapula ventral to the rhomboideus. In quadrupeds it forms, together with the pectoralis, a muscular sling that transfers much of the weight of the body to the pectoral girdle and appendages.

That portion of the serratus ventralis which arises from the cervical vertebrae is at times termed *levator scapulae* although it is not a separate muscle as in man. It draws the scapula ventrally and anteriorly. In humans the *serratus anterior* is homologous to the serratus ventralis.

Scalenus — This is a long muscle that extends longitudinally along the ventro-lateral surface of the neck and thorax. It is divisible into three separate muscles: the *scalenus anterior, medius,* and *posterior.*

The largest and most readily identifiable of the three is the scalenus medius. You can readily identify the scalenus medius and posterior in the photo. The anterior segment is continuous with the *transversus costarum* muscle. The entire scalenus group originates on the ribs and inserts upon the transverse processes of the cervical vertebrae. They bend the neck and draw the ribs anteriorly.

Transversus Costarum — This muscle is not found in humans. It is located near the mid-ventral line where it crosses diagonally from its origin on the sternum to its insertion on the first rib. It acts together with the scalenus. As seen in the photo, it covers the anterior end of the rectus abdominis muscle.

Rectus Abdominis — This muscle was described earlier when the superficial abdominal muscles were discussed. It extends, however, anteriorly to the thorax as well. It originates at the upper ribs and sternum along the mid-ventral line, as seen in the photo. It extends posteriorly, parallel to its partner on the right and left mid-ventral line, to the pubis.

Dorsal View

Supraspinatus and Infraspinatus — The spinous process of the scapula separates two large muscle bundles. The one above the spine, known as the *supraspinatus,* occupies the *supraspinous fossa* of the scapula, while the one below the spine, the *infraspinatus* occupies the *infraspinous fossa.* They lie deep to the acromiotrapezius muscle. They both insert on the greater tubercle of the humerus. The supraspinatus acts to extend the scapula while the infraspinatus rotates the humerus outward.

Teres Major — This is a thick muscle which lies posterior to the infraspinatus. It acts to rotate the humerus and draw it posteriorly. Although this muscle was previously examined during the dissection of the ventral surface, it is quite prominent on both surfaces, covering the posterior border of the scapula and much of its dorsal and lateral surfaces. It inserts in common with the latissimus dorsi on the proximal end of the humerus.

Rhomboideus — Cut and reflect the ends of the spinotrapezius and acromiotrapezius muscles. Three distinct rhomboideus muscles are present. They extend from the upper end of the scapula anteriorly toward the head. They may be seen in the photo, p. 45. They attach the scapula to the vertebral column. The name is derived from their rhomboidal shape.

Rhomboideus Thoracis — This is the most posterior of the three. It originates from the thoracic vertebrae.

Rhomboideus Cervicus — This portion originates from the cervical vertebrae.

Rhomboideus Capitis — This is the most anterior and lateral of the three. It originates from the occipital area, at the back of the skull, to insert on the vertebral border scapula. This portion of the muscle is missing in man.

All three of the rhomboideus muscles act upon the scapula to hold it in place and draw it forward.

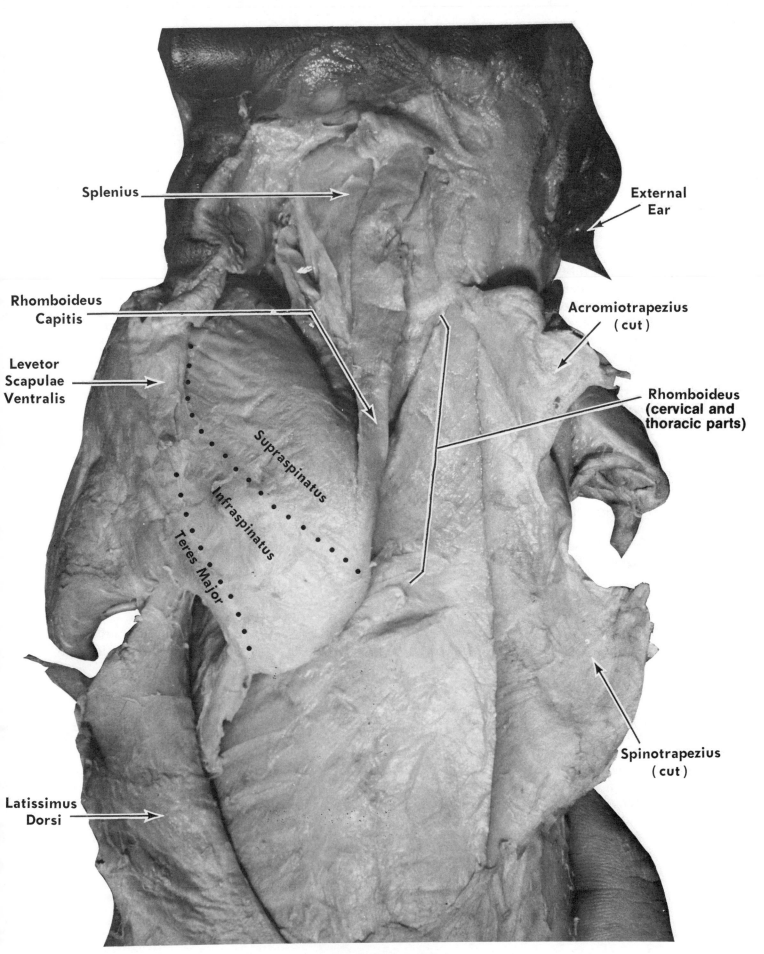

Splenius

External
Ear

Rhomboideus
Capitis

Acromiotrapezius
(cut)

Levetor
Scapulae
Ventralis

Rhomboideus
(cervical and
thoracic parts)

Supraspinatus

Infraspinatus

Teres Major

Spinotrapezius
(cut)

Latissimus
Dorsi

**DEEP MUSCLES
SHOULDER AND NECK, DORSAL VIEW**

DEEP MUSCLES - LUMBAR AND SACRAL AREAS, DORSAL VIEW

Transect and reflect the following muscles:

    spinotrapezius

    latissimus dorsi

In addition, the very prominent *lumbodorsal fascia*, the white aponeurosis covering most of the lumbodorsal surface is cut to reveal the deeper muscles below.

Extensor Dorsi Communis — This large muscle mass on each side of the vertebral column extends from the sacrum and ilium to the skull. As its name indicates, it is an extensor of the spine. It also draws the ribs posteriorly and bends the neck and spinal column to one side. It is comparable to the *sacrospinalis (erector spinae)* of humans.

It is divided longitudinally into three columns of muscle tissue:

Iliocostalis — This is the most lateral of the three. In the thoracic region it inserts upon the ribs and in the lumbar region upon the transverse processes of the vertebrae. It arises from the ilium, the lumbar vertebrae, and the ribs.

Longissimus — This bundle is medial to the iliocostalis. It extends to the skull. Different names for the various segments are used according to their location. These are the *longissimus dorsi, cervicis,* and *capitis.*

Spinalis (Spinalis Dorsi) — This is the most medial of the three muscle columns. It consists of diagonal fibers which insert on the spinous processes of the upper lumbar, thoracic, and cervical vertebrae.

Three other dorsal muscles are:

Splenius — This muscle is located on the back of the neck. It lies deep to the clavotrapezius and rhomboideus muscles. It arises from the first two thoracic vertebrae and from the dorsal midline of the neck to insert upon the occipital bone of the skull. It is an extensor of the head and flexes it laterally.

Multifidus — This muscle extends the entire length of the vertebral column, from the sacrum to the skull, filling the grooves adjacent to the spinous processes of the vertebrae. It may be observed in the lumbar region medial to the longissimus dorsi. It acts to erect and rotate the spine.

Semispinalis — This muscle lies deep to the splenius. It has two sections, the *semispinalis cervicis* and *capitis.* They arise from the vertebrae between the anterior portion of the spinalis. These muscles extend forward to the neck and head.

The deep muscles of the back are all concerned with the extension and lateral flexion of the back, neck, and head. They are called *epaxial,* and are innervated by the smaller dorsal rami of the spinal nerves. The other skeletal muscles of the pig are *hypaxial.*

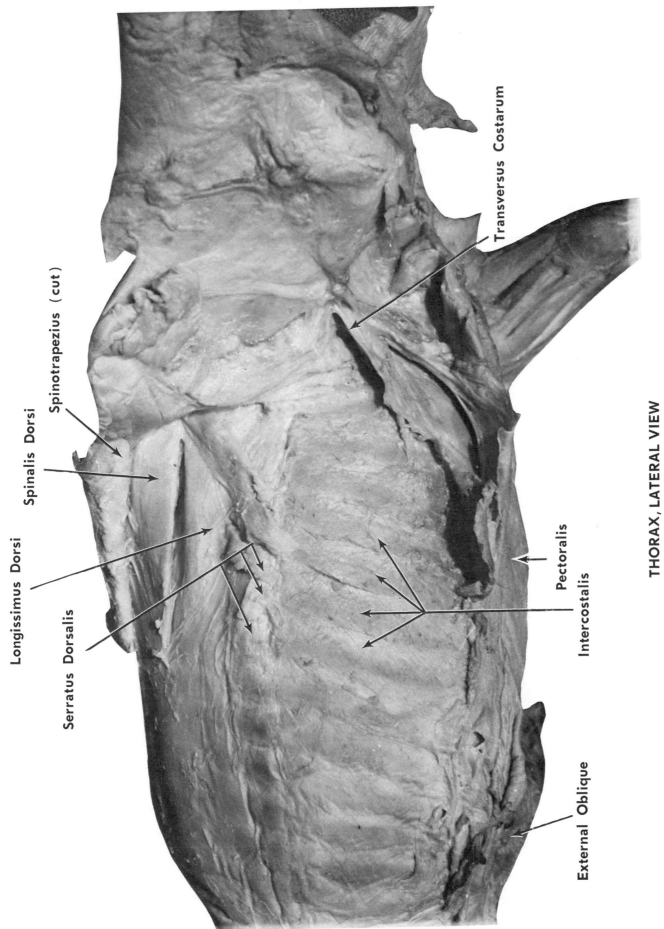

Spinotrapezius (cut)

Spinalis Dorsi

Longissimus Dorsi

Serratus Dorsalis

Transversus Costarum

Pectoralis

Intercostalis

External Oblique

THORAX, LATERAL VIEW

47

Name .... Section .... Date ....

SELF - QUIZ II
MUSCULAR SYSTEM

1. The skin is connected to the muscle layers below by a fibrous tissue known as _____
2. The bone to be moved by a skeletal muscle serves as the:
   (a) origin, (b) insertion, (c) rotation, for that muscle.
3. Name the muscle that acts as the antagonist of the biceps brachii.
4. The ventral and lateral portions of the abdominal wall are composed of three layers of muscle. Name these in the proper order beginning with the outermost one.
5. The major superficial muscle of the upper back and shoulder is the _____
6. Name the tough connective tissue that ties the gastrocnemius muscle to the calcaneus bone.
7. Name the muscles that raise and lower the ribs.
8. Name the large, broad sheet of muscle that originates along the thoracic and lumbar vertebrae and inserts upon the humerus.
9. The sartorius and gracilis are superficial muscles located at the:
   (a) head, (b) chest, (c) arm, (d) leg.
10. Define each of the terms listed below.

ANSWERS

1. _____
2. _____
3. _____
4. _____
5. _____
6. _____
7. _____
8. _____
9. _____
10. a. aponeurosis _____
    b. pronation _____
    c. flexion _____
    d. circumduction _____
    e. linea alba _____
    f. orbicularis oris _____
    g. masseter _____
    h. hyoid _____
    i. synergistic _____

Label all of the features indicated on the photograph.

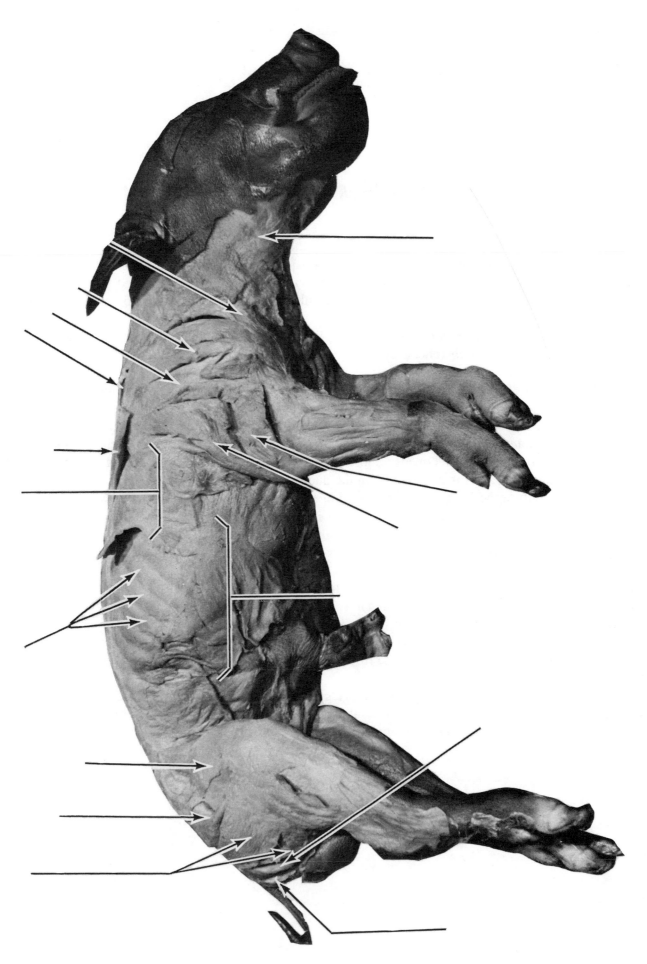

DIGESTIVE SYSTEM - SALIVARY GLANDS

Turn the fetal pig on its side. Carefully cut through the skin along the side of the head between the ear and the mouth, as in the accompanying photo. Separate the skin from the underlying tissues and remove it. The *platysma* and other muscles associated with the skin are also removed. Expose the salivary glands and ducts. This is to be done very carefully, since the structures to be studied lie right below the skin.

In the cheek area note the thick *masseter muscle*. This is the major muscle used in chewing. It originates upon the zygomatic arch (cheek bone), and inserts upon the mandible. When it contracts, it brings the jaws together.

Parotid Gland — This is the largest of the salivary glands. It lies ventral to the pinna (external ear) and is recognized by its triangular shape and lobular texture. The *parotid duct,* or *Stensen's duct,* can be seen emerging from the anterio-ventral edge of the gland by several roots. It then crosses the lower portion of the *masseter muscle* of the cheek, following the course of the prominent *external maxillary vein.* It continues to the mouth and opens opposite the upper fourth molar. You can find the opening by looking inside the cheek and tugging lightly on the duct with forceps.

Note: do not confuse the branches of the *facial nerve* leading to the facial muscles with the parotid duct. The *dorsal buccal branch* and the *ventral buccal branch* of the facial nerve also cross the masseter muscle from beneath the edges of the parotid gland. The parotid duct is generally thicker than the nerve branches. The relationship between the nerve and the parotid duct can clearly be seen in the photo.

Small *buccal glands* lie beneath the skin of the lips.

Submaxillary Gland (*Mandibular Gland*) — Most of this gland lies beneath and ventral to the parotid gland just posterior to the angle of the jaw. It is small and oval in shape. Separate the two glands. Its duct, *Wharton's duct,* is hard to trace since it passes amongst some of the jaw muscles. It extends to the angle of the jaw, passes to the floor of the mouth, to the anterio-ventral connection of the tongue. At this point it opens into the mouth.

Sublingual Gland — This is the third salivary gland. It is hidden in the third salivary photo. It is flat and narrow and lies along the duct of the submaxillary gland. Its duct parallels that of the submaxillary and both open under the tongue, along its side.

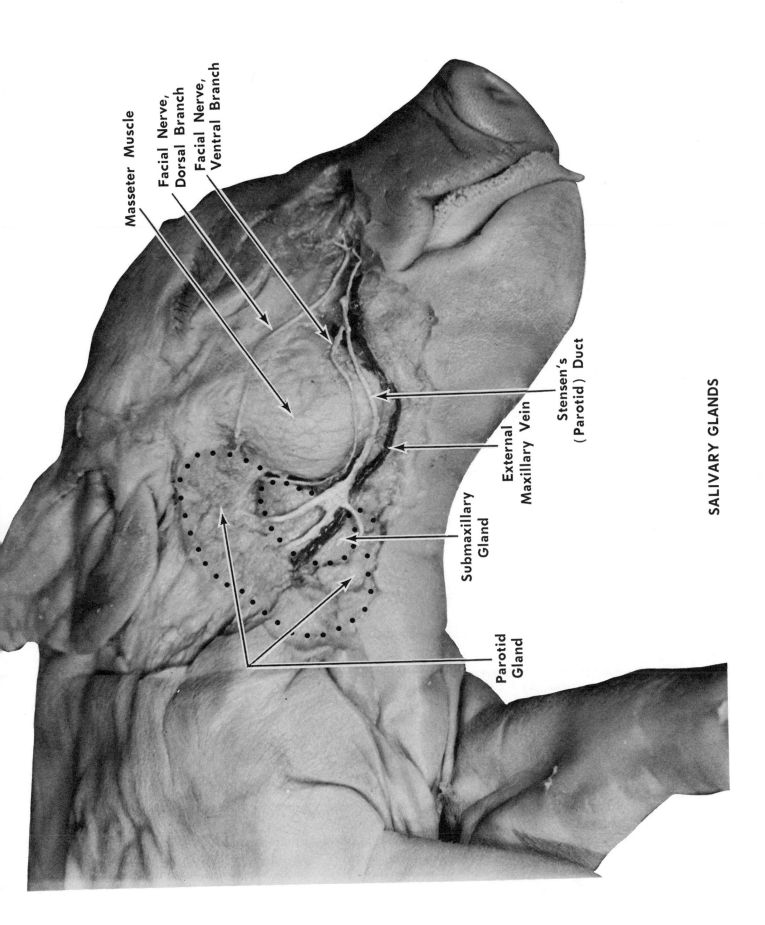

Masseter Muscle

Facial Nerve, Dorsal Branch

Facial Nerve, Ventral Branch

Stensen's (Parotid) Duct

External Maxillary Vein

Submaxillary Gland

Parotid Gland

SALIVARY GLANDS

ORAL CAVITY

With your scissors cut through the *corner of the mouth* on each side in a posterior direction. Continue cutting through the *angle of the jaw*. Expose the entire tongue. The interior of the oral cavity may now be examined.

Vestibule — This refers to the area between the lips and the teeth.

Tongue — This elongated muscular structure is readily visible upon the floor of the mouth. It is attached vertically along much of its length by a membrane, the *lingual frenulum*, and posteriorly to the *hyoid* bone. The surface of the tongue is covered by variously shaped projections known as sensory *papillae*. The greatest number of large fibrous papillae are to be seen at the anterior edge of the tongue. Microscopic *taste buds* are found at the sides and base of the papillae.

Teeth — Upon the upper jaw two *canine* teeth are visible in the photo, one on each side. These, and the third pair of *incisors* are the first to erupt.

The dental formula of the fetal (young) pig is: $I\frac{3}{3}, C\frac{1}{1}, P\frac{4}{4}, M\frac{0}{0}$.

The adult pig: $I\frac{3}{3}, C\frac{1}{1}, P\frac{4}{4}, M\frac{3}{3}$.

Compare this to the dental formula of man. *Deciduous* or first teeth of humans: $I\frac{2}{2}, C\frac{1}{1}, P\frac{2}{2}, M\frac{0}{0}$.

The adult human: $I\frac{2}{2}, C\frac{1}{1}, P\frac{2}{2}, M\frac{3}{3}$.

In the young of both pig and man the *molars*, the large, broad grinding teeth have not yet erupted. Three of these will appear in each quarter of the adult pig, and three in the human.

Both the pig and man are *omnivores* whose diet consists of both plant and animal sources. The types of teeth of mammals are indicative of their mode of nutrition.

The pig's dental pattern as well as that of man are characteristic of an *omnivorous* diet. The cat and dog are *carnivorous*. Their teeth are sharp and pointed, fewer molars and modified premolars. Horses, cows and other *herbivorous* animals possess mainly large, broad, flat surfaced molars for grinding; plus a double row of incisors at the front of the mouth for cutting and shearing vegetation.

Palate: This structure forms the roof of the mouth. It is a partition which separates the oral from the nasal cavity.

> **Hard Palate:** This is the bony anterior portion of the palate. A series of transverse ridges, the *palatine rugae*, cross the roof of the mouth.
>
> **Soft Palate:** This is the posterior continuation of the palate. It is a muscular structure with bony support. It divides the *oropharynx* ventrally from the *nasopharynx* dorsally. In man there is a finger-like process, the *uvula*, which hangs down from its center posteriorly. It is absent in the pig.

Slit the soft palate longitudinally and observe the nasopharynx. The *Eustachian tubes* pass from the latero-dorsal wall of the nasopharynx to the middle ear. At the anterior end you will find the openings of the *internal nares*. They are continuous with the *external nares,* or nostrils.

Near the entrance to the nasopharynx find the *isthmus of fauces,* the opening from the oral cavity into the oropharynx.

Epiglottis — This cone-shaped flap of cartilage is located at the top of the *larynx* (voice box) near the base of the tongue. It protects the *glottis,* the slit-like opening to the trachea. During swallowing and eating the epiglottis prevents food from entering the trachea.

Trachea — This tube is commonly called the windpipe. It is topped by the epiglottis and larynx. It is kept open by rings of carfilage which extend around the trachea at intervals. They are incomplete dorsally. The trachea branches to form two *bronchi* which enter the lungs.

Esophagus — This muscular tube, located dorsal to the trachea is also known as the gullet. Unlike the trachea, however, it is collapsed. Food is pushed forward in the esophagus by the rhythmic contractions of its walls, a process knows as *peristalsis.*

The esophagus extends posteriorly and dorsally within the thorax, then passes through the diaphragm into the abdominal cavity where it ends at the stomach.

In order to find the trachea and esophagus use two wooden probes. With one, penetrate the glottis and pass into the trachea. Move the probe up and down and observe the movement of the trachea. With the second probe enter the esophagus dorsal to the glottis. Move it up and down and observe the corresponding movement of the esophagus (see photo p. 55).

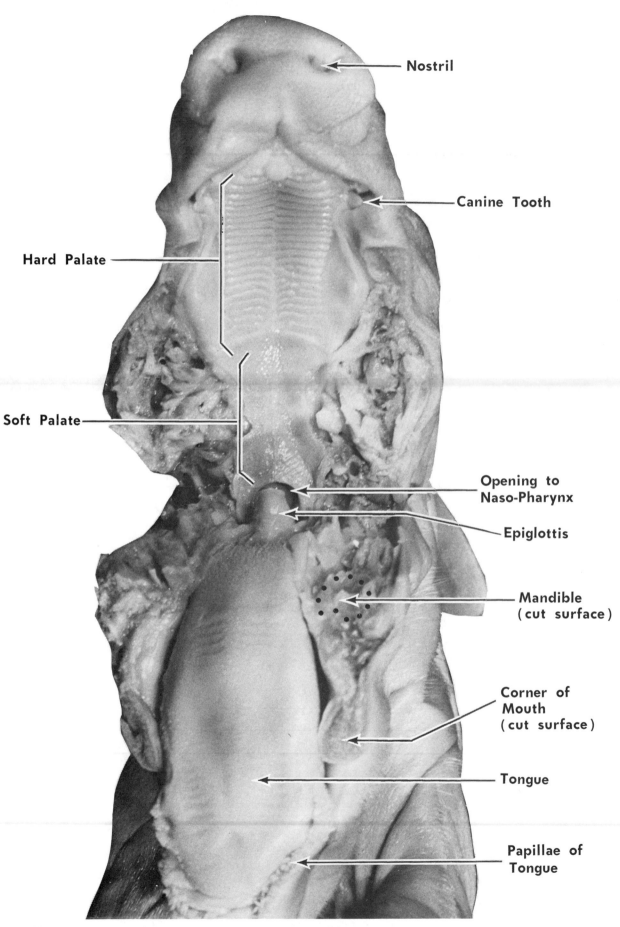

Nostril

Canine Tooth

Hard Palate

Soft Palate

Opening to
Naso-Pharynx

Epiglottis

Mandible
(cut surface)

Corner of
Mouth
(cut surface)

Tongue

Papillae of
Tongue

ORAL CAVITY

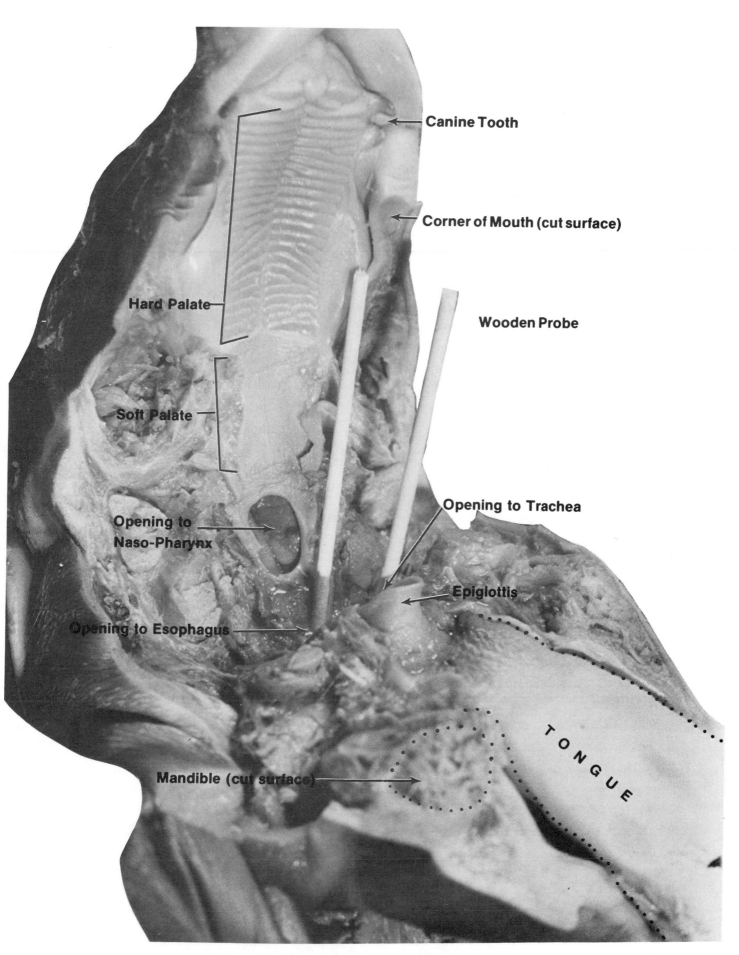

Canine Tooth

Corner of Mouth (cut surface)

Hard Palate

Soft Palate

Wooden Probe

Opening to
Naso-Pharynx

Opening to Trachea

Epiglottis

Opening to Esophagus

Mandible (cut surface)

T O N G U E

THE ORAL CAVITY (CLOSE-UP)

SELF - QUIZ III
THE ORAL CAVITY

1. Name the divisions of the pharynx.
2. Name seven passageways that penetrate the pharynx.
3. Name four different types of papillae found on the surface of the tongue. How do they differ?
4. Name the bones that comprise the hard palate.
5. Besides the palatine tonsils, what other tonsils do you possess?
6. The secretions of the salivary glands begin the digestion of, a) proteins, b) sugars, c) starches, d) lipids.
7. The esophagus is located, a) to the right of the trachea, b) to the left of the trachea, c) dorsal to the trachea, d) ventral to the trachea.
8. A major blood vessel is formed just posterior to the submaxillary gland as a result of the union of smaller vessels. Name the major blood vessel formed.
9. Name the cap of cartilage which prevents food from entering the trachea while swallowing.
10. Define each of the terms listed below.

ANSWERS

1. _____
2. _____
3. _____
4. _____
5. _____
6. _____
7. _____
8. _____
9. _____
10. a. dentition _____
   b. eustachian tube _____
   c. pharynx _____
   d. larynx _____
   e. naso-pharynx _____
   f. frenulum linguae _____
   g. deciduous teeth _____
   h. glottis _____
   i. Stensen's duct _____
   j. Wharton's duct _____

Label all of the structures indicated on the photograph.

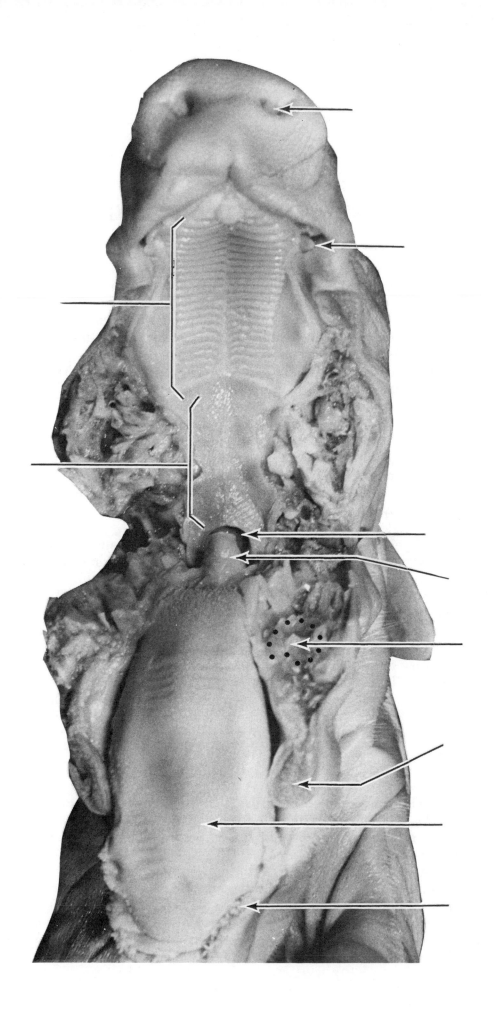

THE ABDOMINAL CAVITY

The muscular *diaphragm* separates the upper from the lower ventral body cavity. The upper is the *thoracic*, the lower is the *abdominal cavity*. We shall study the abdominal area first and later consider the thorax in relation to the study of the heart and circulatory system.

With your fingertips locate the lower edges of the ribs. Your fingertips will be tracing an arc, an inverted letter "V". Refer to the photo entitled, "Mapping Incisions," p. 61. Make the cuts in the order of the numbers indicated, beginning with No. 1. Do not make incisions No. 5 and No. 6 until you have completed the observations of the abdominal viscera and you are ready to observe the *thoracic organs*. This will prevent the thoracic area from drying out prematurely.

Use your scalpel to cut the musculature along the line you have traced with your fingertips and indicated as No. 1 on the photo. Do not cut too deeply. The skin and muscles of the fetal pig are very thin and soft. A sharp scalpel in an untrained hand may lead to the destruction of internal organs and possible injury to the student.

Continue with incision No. 2. This will bring you just above the *umbilical cord*. Cut around the cord (incision No. 3) to avoid injury to the *umbilical arteries*, *umbilical vein*, the *urinary bladder* and the *penis* (in males). Extend incision No. 4 to the rear body wall.

After the muscle layers have been cut you will find a fine membrane, the *peritoneum*, which lines the inside of the abdominal cavity. The portion of this serous membrane that you see is the *parietal peritoneum*, the *visceral peritoneum* covers the abdominal viscera. Cut through the peritoneum, fold back the entire *ventral abdominal wall* to expose the organs below. You will note that the muscular wall below the *umbilical cord* cannot be lifted. This is because of the *umbilical vein* which passes into the liver. It is necessary to cut the vein at this time to expose the abdominal cavity.

Some specimens may contain excess preservative fluid, coagulated blood, or dye that has escaped from the blood vessels. In these cases it is first necessary to wash out the abdominal cavity. Hold the pig under a moderate flow in the sink and rinse gently. Use paper towels to soak up excess water. Your view should now correspond to that in the photos on page 62 and to the close-up on page 63.

(Note: in both photos we observe the abdominal cavity as it appears when we first begin the dissection. Some of the "hidden" structures are not labeled in the photographs. It will be necessary to move other organs from their natural positions in order to expose them.)

Identify the following structures:

Diaphragm — This dome-shaped muscular wall separates the thoracic from the abdominal cavity. It is also the most important muscle for *breathing*, permitting inhalation and exhalation. Three major vessels pass through the diaphragm between the thorax and the abdomen. These are the *aorta, posterior vena cava,* and the *esophagus.*

Liver — This dark brown organ dominates the upper abdomen. The *falciform ligament*, a ventral peritoneal membrane attaches the liver to the diaphragm and to the ventral body wall. The *coronary ligament* attaches the dorsal portion of the liver to the central tendon of the diaphragm.

The falciform ligament lies in a cleft of the liver which divides it into right and left halves. Five lobes can be differentiated. The four principal lobes may be seen from the ventral aspect, they are the *right lateral, right central, left central,* and *left lateral.* A very small lobe, the *caudate lobe,* may be seen when the intestinal coils are moved to the left. It is attached to the posterior surface of the right lateral lobe.

Gall Bladder — Lift the right central lobe of the liver and expose the gall bladder embedded within a depression in its dorsal surface. This sac-like structure stores bile secreted by the liver and releases it into the duodenum. Bile is transported by the *cystic duct* from the gall bladder. It is joined by the *hepatic duct* from the liver to form the *common bile duct* which enters the duodenum. These can be clearly seen in the accompanying photo, p. 64.

Stomach — This muscular pouch lies on the left side in the upper abdomen. It is the continuation of the esophagus. Find the esophagus and locate where it pierces the diaphragm to join the stomach. This is the *cardiac* end of the stomach. The *fundus* is the dilated anterior portion, the *body* is the main portion, while the *pyloric* region is the most posterior. This end joins the duodenum.

Open the stomach with your scissors by cutting along the *greater curvature* of the stomach, on the left side. Wash out the contents of the stomach. Note the *cardiac sphincter* which controls the entrance of food into the stomach from the esophagus. The *pyloric sphincter* at the posterior end regulates the release of partially digested food (chyme) into the duodenum. Look along the inner walls of the stomach and note the *rugae*, or folds which help to churn and mix the food with digestive juices.

The green debris found in the stomach and elsewhere in the digestive tract is called *meconium*. Since the animal is still in the fetal state it does not represent food actually eaten. It is a combination of bile-stained mucus, epithelial cells sloughed off from the skin and lining of the digestive tract, and amniotic fluid swallowed by the fetus. It will be discharged in the first bowel movement of the newborn.

Small Intestine — The first portion of the small intestine is the *duodenum*. It is a continuation of the pyloric end of the stomach. It is a short "U" shaped tube, approximately 1 cm. long. The common bile duct and the pancreatic duct open into the duodenum. The second section of the small intestine is the *jejunum*, which makes up about half the length of this organ. The *ileum* is the final section. Open the jejunum or ileum, wash its contents and touch the inner surface with your fingertips. The velvety texture felt is due to the presence of numerous *villi* along the inner walls. Use a hand lens or a low power dissection microscope to observe them more clearly.

The coils of the small intestine are held in place by a fine peritoneal membrane, the *mesentery*. It may be observed when lifting a coil of the small intestine and stretching the two ends. A fine, thin membrane, the mesentery, will be visible. It is responsible for the coiling observed. Note its shiny thin appearance. It is interlaced with narrow blood vessels, lymphatic vessels, adipose tissue, and lymph nodes. Some of the tiny blood vessels form the beginnings of the portal system, transporting digested food from the intestine to the liver. Cut through the mesentery to unravel the small intestine. Measure its length. How does it compare to the relative length of man's intestine (about twenty feet)?

Large Intestine — Follow the coils of the small intestine. The end of the ileum enters the large intestine. At this juncture the *caecum*, a short blind sac about 2 cm. long, is formed. In some organisms such as horses, this section is enlarged and houses microorganisms which can digest cellulose. Humans possess a *vermiform appendix* that projects from the end of a short caecum. Cut into the caecum at about the point where the ileum enters. Wash out its contents, look for and locate the *ileocaecal* valve.

The *spiral colon* is a compact coiled mass clearly visible upon the left ventral surface. It is shorter, darker, and thicker than the small intestine. It is the major portion of the large intestine. The posterior dorsal portion of the large intestine is the *rectum*. It descends along the midline through the pelvic girdle to the *anus*, the intestinal opening to the exterior. The colon of human beings is relatively shorter than that of the fetal pig and is not coiled.

Pancreas — Lift the main portion of the small intestine. Expose the stomach and duodenum. Observe the pancreas, a lobulated glandular structure lighter in color than the neighboring intestines. Its main portion lies in the loop of the duodenum. An elongated portion may be observed extending to the left, toward the stomach and spleen. Parts of the gland may also be seen along the dorsal body wall extending to the right of the duodenum and along the dorsal midline. The human pancreas is much more compact. Its duct, the *pancreatic duct*, enters the duodenum. It is small and difficult to find since it is embedded in glandular material.

Spleen — This dark-colored elongated organ can readily be seen in the left side of the abdominal cavity without moving any other organs. It lies to the left of the stomach, along its greater curvature, and extends toward the right. It is tied to the stomach by a portion of the *greater omentum*, a specialized fold of the periotoneum, known as the *gastrosplenic ligament.*

Kidney — This reddish-brown bean-shaped organ lies embedded retroperitoneally, namely, behind the parietal peritoneum, in the dorsal body wall, one on each side. The *adrenal gland* is located near the anterior end of the kidney, but is separated from it, lying slightly mediad of the kidney. In humans the adrenal gland forms a "cap" upon the kidney.

Urinary Bladder — In the fetal pig the urinary bladder is an elongated sac in the lower ventral abdominal cavity. It lies between the prominent *umbilical arteries* and is seen when the portion of the body wall supporting the umbilical cord is folded down.

Reproductive Structures — Most of the female reproductive structures and some of the male's are located in the abdominal cavity. The urogenital system and its associated structures will be studied and more fully discussed in a later chapter.

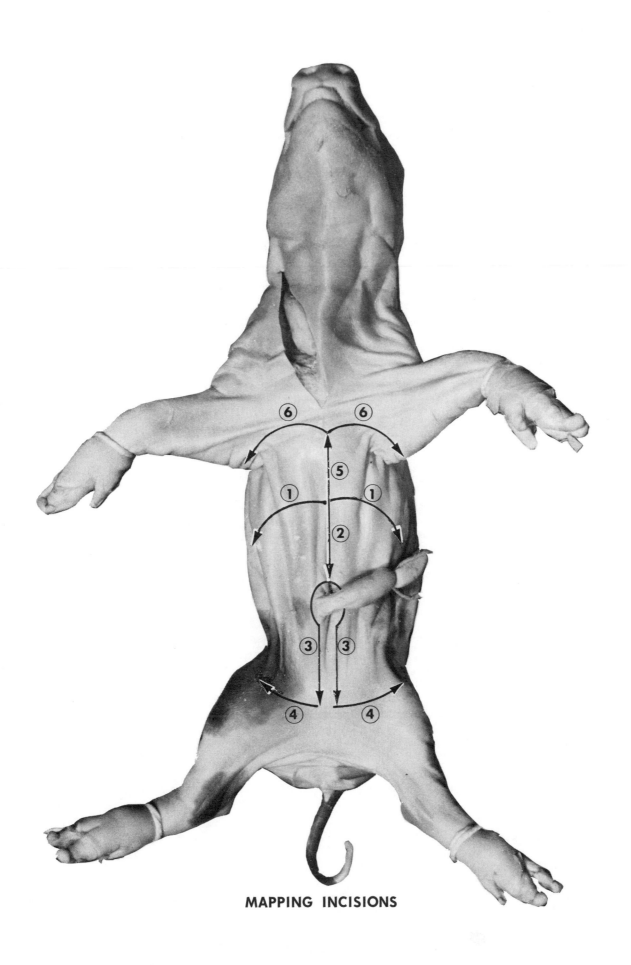

MAPPING INCISIONS

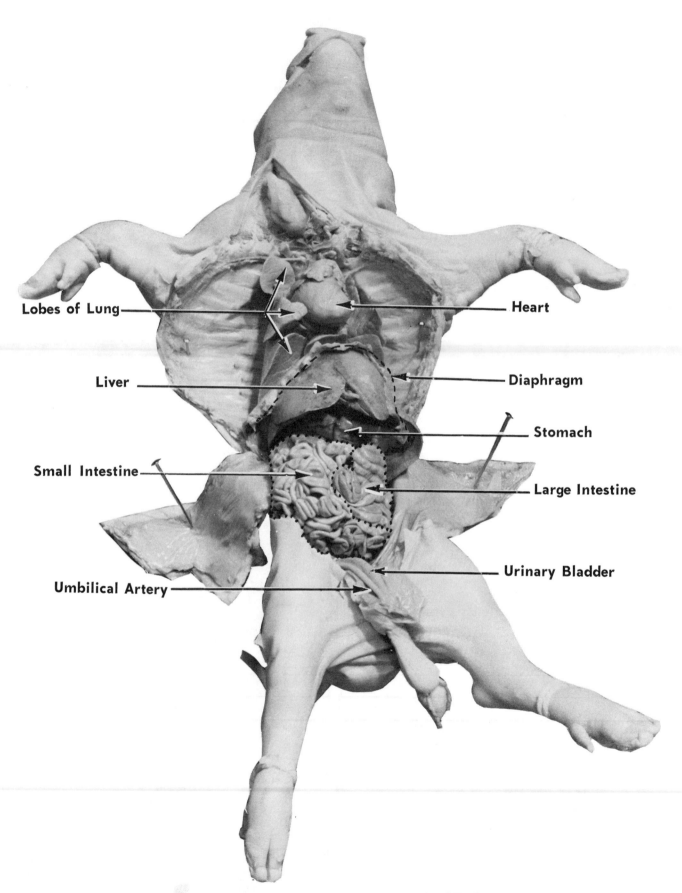

Lobes of Lung

Heart

Liver

Diaphragm

Stomach

Small Intestine

Large Intestine

Urinary Bladder

Umbilical Artery

THORACIC AND ABDOMINAL VISCERA

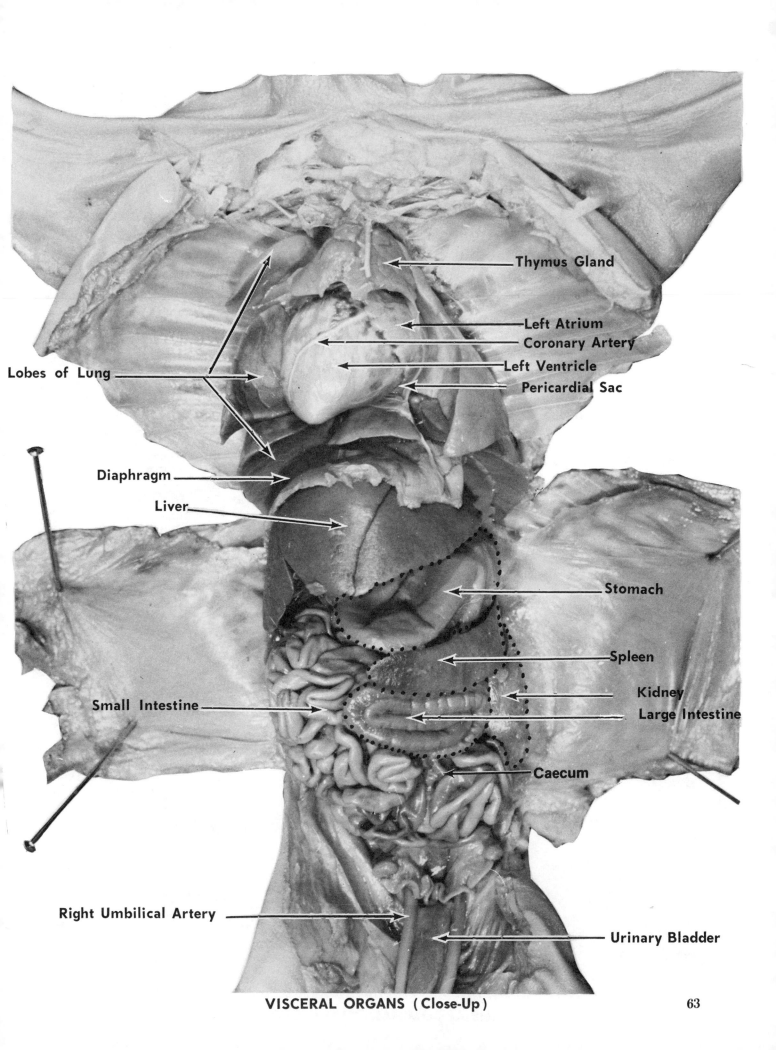

Thymus Gland

Left Atrium

Coronary Artery

Left Ventricle

Pericardial Sac

Lobes of Lung

Diaphragm

Liver

Stomach

Spleen

Kidney

Large Intestine

Small Intestine

Caecum

Right Umbilical Artery

Urinary Bladder

VISCERAL ORGANS (Close-Up)

63

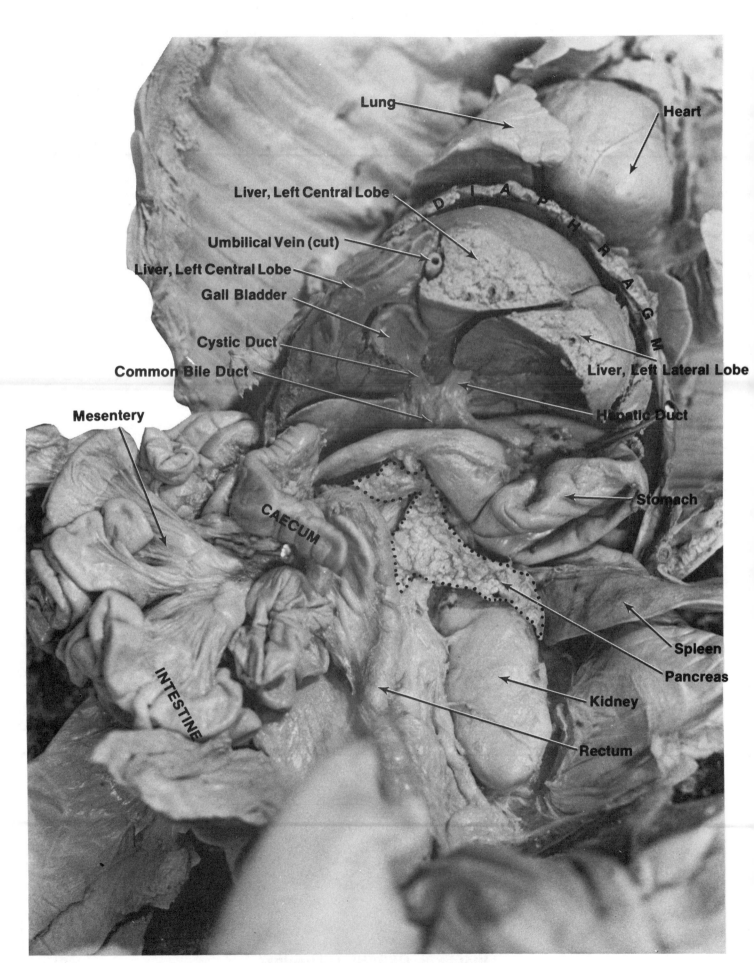

Lung

Heart

Liver, Left Central Lobe

D I A P H R A G M

Umbilical Vein (cut)

Liver, Left Central Lobe

Gall Bladder

Liver, Left Lateral Lobe

Cystic Duct

Common Bile Duct

Hepatic Duct

Mesentery

Stomach

CAECUM

INTESTINE

Spleen

Pancreas

Kidney

Rectum

THE ABDOMINAL CAVITY (CLOSE-UP)

HUMAN DIGESTIVE SYSTEM

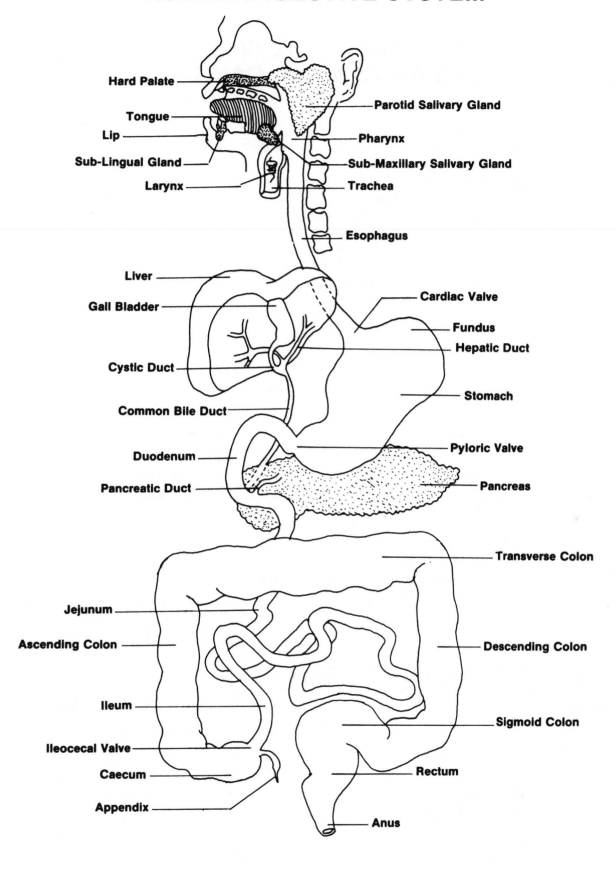

Hard Palate

Tongue

Lip

Sub-Lingual Gland

Larynx

Parotid Salivary Gland

Pharynx

Sub-Maxillary Salivary Gland

Trachea

Esophagus

Liver

Gall Bladder

Cystic Duct

Common Bile Duct

Cardiac Valve

Fundus

Hepatic Duct

Stomach

Duodenum

Pancreatic Duct

Pyloric Valve

Pancreas

Transverse Colon

Jejunum

Ascending Colon

Ileum

Ileocecal Valve

Caecum

Appendix

Descending Colon

Sigmoid Colon

Rectum

Anus

SELF - QUIZ IV
THE ABDOMINAL CAVITY

1. Name the five lobes of the liver.
2. Describe the structures of the greater and lesser omentum.
3. How do the jejunum and ileum differ?
4. What is the location and function of the ileocecal valve?
5. Name the parts of the alimentary canal in their proper order.
6. How does the location of the adrenal gland differ in the pig and in man?
7. The digestion of fats begins in the, a) mouth, b) stomach, c) small intestine.
8. Name the three major tubular structures that pass through the diaphragm?
9. Which of the following is not found in the abdominal cavity?
   a) gall bladder, b) kidney, c) trachea, d) pancreas

ANSWERS

1. _____
2. _____
3. _____
4. _____
5. _____
6. _____
7. _____
8. _____
9. _____

10. Define, identify, and locate each of the structures listed below.

   a. esophagus _____
   b. duodenum _____
   c. jejunum _____
   d. colon _____
   e. fundus _____
   f. rugae _____
   g. pyloric sphincter _____
   h. mesentery _____
   i. cystic duct _____
   j. common bile duct _____

Label all of the features indicated on the photograph.

RESPIRATORY SYSTEM - THE THORACIC CAVITY

Begin your dissection of the thoracic cavity by making an incision with your scissors at the midventral base of the rib cage. Follow incision No. 5 as in the photo on p. 61. Continue your incision until the top rib has been cut. Then cut laterally as incision No. 6. Separate the edges of the *diaphragm* from the ventral and lateral walls of the thorax. You are now ready to spread the rib cage and expose the *heart* and *lungs.*

Next, cut anteriorly along the mid-ventral line into the musculature of the neck toward the chin. Separate the neck muscles to expose the *trachea, larynx, thyroid gland, jugular veins,* and *carotid arteries,* as in the accompanying photo.

Note the following:

Thymus Gland — This whitish glandular body partially covers the heart. Two major lobes extend anteriorly into the neck region on either side of the trachea. It is necessary to remove the lower portion in order to study the heart. This gland is enlarged in the fetus and in younger animals, then becomes reduced as the animal matures. Only a small portion remains in the adult.

Pericardium — This fibrous, double-layered membranous sac encloses the heart and the large blood vessels at the anterior end of the heart. Remove it in order to expose the heart. The *phrenic nerves* which innervate the diaphragm pass along the lateral edges of the pericardium. Identify these nerves.

Heart — This conical organ is located in the center of the thorax, within the *mediastinum,* the space between the lungs. It consists of two *atria* and two *ventricles.* A detailed study of the heart will be made when the circulatory system is studied. Note the *coronary arteries* and *veins* on the surface of the heart. They supply the heart muscle itself with blood.

The large *pulmonary artery* is seen leaving the heart from its ventral surface extending toward the left side. More distally the *aortic arch* leaves the heart and also extends toward the left side. In the fetal pig the pulmonary artery is joined directly to the aortic arch by means of a short vessel, known as the *ductus arteriosus.* It serves as a bypass to shunt the blood from the lungs into the systemic circulation. This connecting link persists till birth. It then shuts tightly, separating the two major blood vessels. It persists in the adult as only a small tendinous band.

Push the heart gently to the left. Note a wide, blue blood vessel rising anteriorly from the diaphragm mid-dorsally. This is the *posterior vena cava.* In man, due to his upright position, this vessel is known as the *inferior vena cava.* It returns blood from the lower portion of the body and enters the heart at the right atrium. A similar, but shorter, blood vessel is seen anterior to the heart. This is the *anterior vena cava,* or *superior vena cava,* in man. It returns blood from the upper positions of the body, from the head and forelimbs. It also enters the heart at the right atrium.

Lungs — Examine the pink lungs on either side of the heart. Note that the lungs of the fetus are firmer than the more spongy lung tissue found after birth. The larger, right lung is divided into four lobes; the *apical, cardiac, diaphragmatic,* and a fourth smaller lobe below the apex of the heart, the *intermediate.* The left lung is divided into three lobes. The small fourth lobe is missing. In humans the right lung has three lobes, the left lung only two. Each lung lies within a separate *pleural cavity,* the space between the lung and the thoracic body wall.

Remove a small, thin section of lung and observe with a hand lens or low power dissection microscope. Note that the fetal lungs are filled with fluid not air. Will the lung float in water? If your specimen has been double injected (arteries and veins) you should observe three types of vessels within the lung tissue:

1. **Pulmonary Artery** — Branches of this vessel contain *blue* dye.
2. **Pulmonary Vein** — Branches of this blood vessel contain *red* dye.
3. **Bronchioles** — These branches of the bronchi, distributed throughout the lungs, are hollow with *white*-edged walls.

Pleura — This is the serous membrane found within the thorax. The *parietal pleura* lines the inner walls while the *visceral pleura* covers the organs of the thorax.

Trachea — The windpipe, or trachea, is a banded tube which extends along the mid-ventral portion of the neck into the thoracic cavity. Here it branches to form the two *bronchi* which penetrate the lungs. The air passage is always kept open by cartilage rings along its entire length. They give support and shape to the cylindrical walls of the trachea.

Remove a half inch section of the trachea. Examine its structure. Cut it lengthwise across the rings. Note that the cartilage rings are incomplete dorsally, thus forming the letter "C." Observe the inner surface with a band lens.

Esophagus — The food pipe, or esophagus, lies dorsal to the trachea and extends through the thorax along the left side. Move the left lung toward the midline and examine the muscular esophages below. Do not mistake it for the thoracic *aorta* which lies along the dorsal midline. It passes through the diaphragm into the abdominal cavity to join the stomach.

Larynx — This structure is also known as the voice box. It is located at the top of the trachea. Its uppermost segment is the triangular flap of tissue, the *epiglottis,* which protects the opening to the trachea.

Cut into the larynx with your scalpel along the mid-ventral line and separate the right from the left side. Examine the inner surface of the larynx. Locate the *vocal folds,* two shelf-like membranes. These are better developed in man as the *vocal cords.* The pitch of your voice depends upon the length, thickness and elasticity of the vocal cords. Changes in pitch are produced by muscles attached to the vocal cords which can alter the tension with which they are held.

Thyroid Gland — This dark oval-shaped gland is located above the trachea just above the rib cage. It is richly supplied with blood vessels.

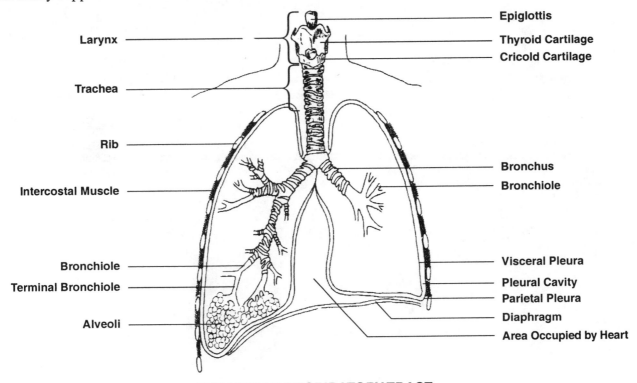

THE HUMAN RESPIRATORY TRACT

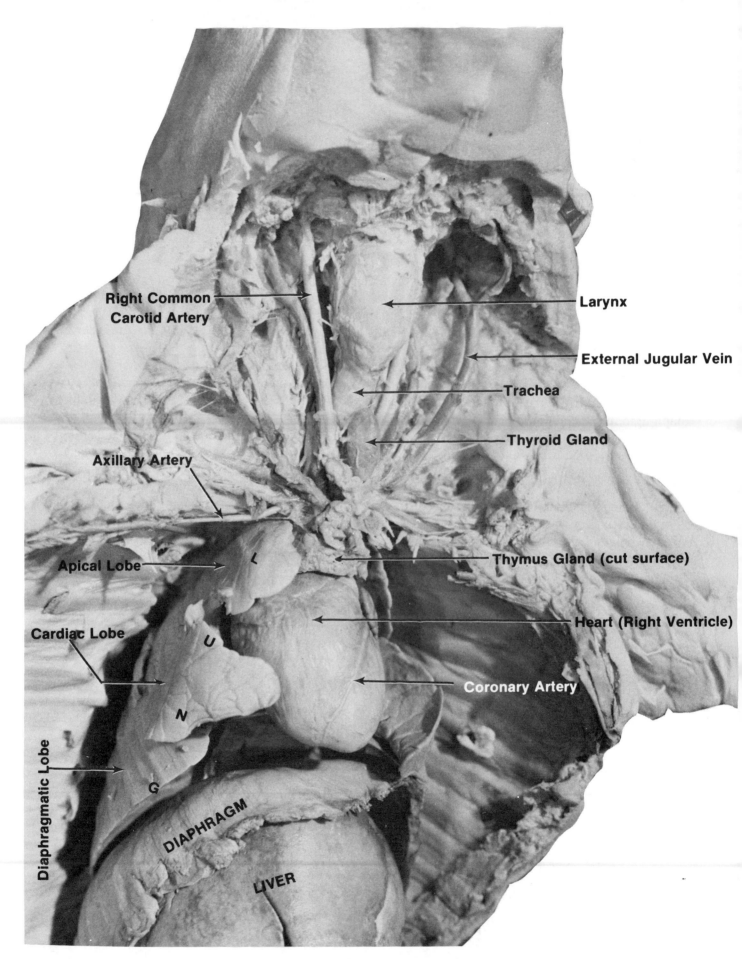

Right Common Carotid Artery

Larynx

External Jugular Vein

Trachea

Thyroid Gland

Axillary Artery

Apical Lobe

L

Thymus Gland (cut surface)

Cardiac Lobe

U

Heart (Right Ventricle)

N

Coronary Artery

Diaphragmatic Lobe

G

DIAPHRAGM

LIVER

THE THORACIC CAVITY

SELF -QUIZ V
THE THORACIC CAVITY

1. Name the lobes of the pig's lungs.
2. How many pairs of ribs does the pig possess? man?
3. What three types of vessels are seen when we section the lung?
4. Describe the structure of the pericardium.
5. Do the bronchi contain cartilage rings? the bronchioles? the alveoli?
6. We inhale 20% oxygen and 0.04% carbon dioxide. What percentages of the exhaled air are oxygen and carbon dioxide?
7. Name the cartilaginous structures that make up the larynx.
8. Name the hormones secreted by the thyroid and their functions.
9. What is the function of the thymus gland?
10. Define each of the terms listed below.

ANSWERS

1. _____
2. _____
3. _____
4. _____
5. _____
6. _____
7. _____
8. _____
9. _____

10. a. tidal volume _____
    b. emphysema _____
    c. pneumonia _____
    d. surfactant _____
    e. mediastinum _____
    f. phrenic nerve _____
    g. Hering-Breuer reflex _____
    h. negative pressure breathing _____
    i. pleurisy _____
    j. parathyroid glands _____

Label all of the features indicated on the photograph.

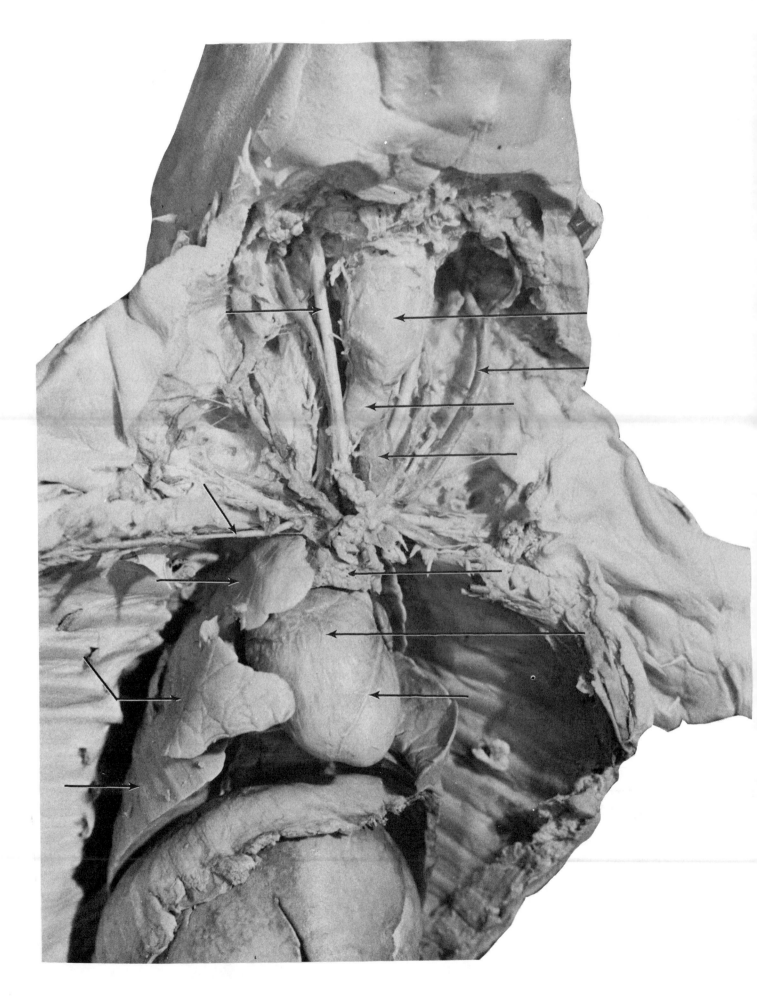

CIRCULATORY SYSTEM - INTRODUCTION

The function of the circulatory system is the *transport* of materials to and from the cells. The organs of multicellular animals are too far removed from the external environment to enable them to exchange nutrients, oxygen and wastes directly by diffusion. Instead, the materials needed by cells must be brought to the cells by a circulatory system, and the cellular wastes must similarly be removed.

The circulatory system includes the circulating medium of transport, the blood. The structures comprising the circulatory system include the:

> **heart** — A muscular pump, responsible for the movement of blood.
>
> **arteries** — Vessels which carry blood away from the heart.
>
> **capillaries** — They connect arteries to veins. The narrowest blood vessels.
>
> **veins** — Vessels which carry blood towards the heart.

You have already examined some of the organs of the circulatory system in the last chapter during the dissection of the thoracic cavity. These included the *heart,* the *pericardium,* and some of the large blood vessels entering and leaving the heart, the *aorta, pulmonary artery,* the *anterior* and *posterior vena cava.*

The fetal circulation has two adaptations for bypassing the inactive lungs. The first is the *ductus arteriosus* which was already described on p. 68. It is a short vessel which connects the pulmonary artery to the aorta. The second is the *foramen ovale,* a small opening in the septum between the right and left atria which permits blood to enter the left atrium without first going to the lungs.

We shall begin the study of circulation with the *sheep heart.* Its large size enables you to study the structure of the heart and associated blood vessels in great detail.

Next, the venous system of the fetal pig is dissected. Veins generally lie closer to the surface than arteries. In doubly injected specimens they are colored blue. Arteries, injected with red latex, are studied last.

SHEEP HEART, VENTRAL VIEW

The sheep heart is studied because it is larger than the fetal pig's heart and closer in size to that of the human.

Examine the sheep heart and note the conical shape. The tip of the heart, or the apex, is the most posterior section. Much of the fat surrounding the blood vessels at the anterior end of the heart has been removed to facilitate observation. Almost all of the *parietal pericardium*, the fibrous serous membrane enclosing the heart has been removed. In some specimens parts of it will still be found fused to the bases of the large blood vessels. The *epicardium*, or outer layer of the heart, is covered by the finer visceral pericardium.

Hold the heart upright as in the photo, with most of the large vessels opening to the rear. You are now looking at the ventral side of the heart. The right side of the heart is at your left hand; the left side of the heart is at your right hand.

Find and identify the ear-like structures atop the ventricles. These are the *right* and *left atria*. (The term auricle is sometimes used. It means "little ear".)

The *pulmonary artery* is the prominent blood vessel leaving the heart at the right ventricle in the upper mid-ventral area. It passes anteriorly toward the left.

The largest and widest artery of the body, the *aorta,* can be seen next to its first major branch, the *brachiocephalic* artery. This vessel carries blood to the right shoulder and arm as well as to the head.

One of the *coronary arteries* may be seen along the ventral surface of the ventricle. It lies within the *longitudinal sulcus,* a depression upon the surface corresponding to the line of separation between the right and left ventricles. The *coronary veins* generally run along with the arteries.

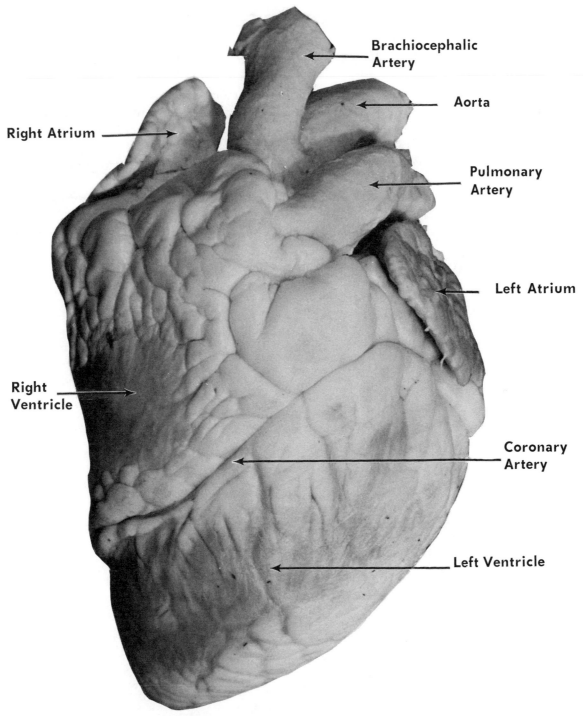

Brachiocephalic Artery

Aorta

Right Atrium

Pulmonary Artery

Left Atrium

Right Ventricle

Coronary Artery

Left Ventricle

SHEEP HEART, VENTRAL VIEW

SHEEP HEART, DORSAL VIEW

Now turn the heart around and observe the dorsal surface. This time the right and left sides of the heart correspond to your own.

Find the *right atrium* and *left atrium*. The *pulmonary* artery which we saw in the last photo is seen here divided into two smaller tubes, the *right* and *left branches*. One passes to each lung.

The wide and thick-walled *aorta* may also be seen along with the *brachiocephalic artery*.

Not visible in the previous photo but clearly seen here are:

— The *superior vena cava* which empties into the right atrium carrying deoxygenated systemic blood from the head, arms, and upper chest.

— The opening of the *inferior vena cava* which carries deoxygenated systemic blood from lower parts of the body.

— the entrance of one of the *pulmonary veins* into the left atrium carrying oxygenated blood from the lungs.

Find each of these structures in your specimen.

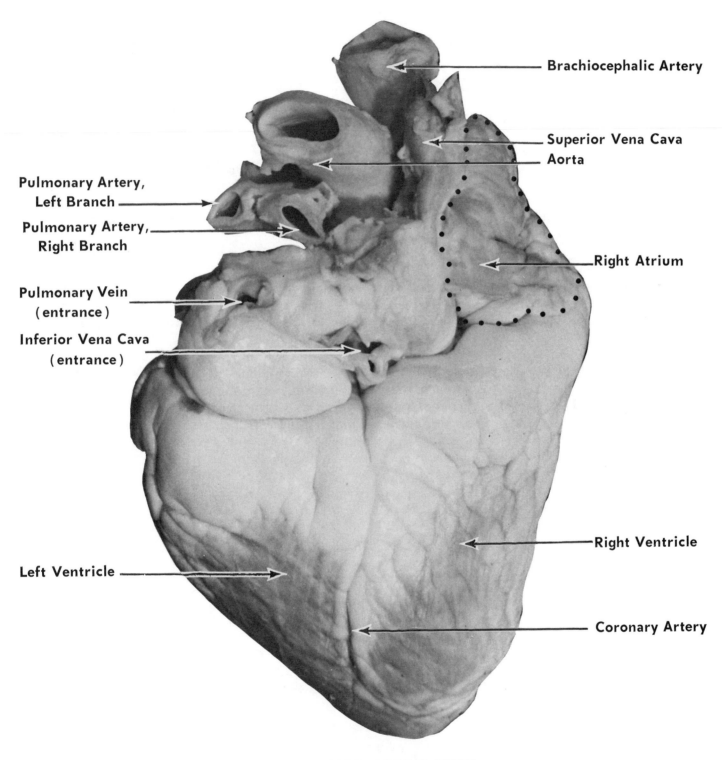

Brachiocephalic Artery

Superior Vena Cava
Aorta

Pulmonary Artery,
Left Branch

Pulmonary Artery,
Right Branch

Right Atrium

Pulmonary Vein
(entrance)

Inferior Vena Cava
(entrance)

Right Ventricle

Left Ventricle

Coronary Artery

SHEEP HEART, DORSAL VIEW

SHEEP HEART (OPEN), CORONAL PLANE

Note: At the start of the sheep heart dissection divide the class in two. Half will follow the dissection procedures outlined here, the other half will dissect as in the following two photos. The groups should then exchange specimens with one another.

Place the sheep heart in a dissection pan, ventral surface up. Use your scalpel to cut into the *myocardium*, the muscle layer that comprises the major portion of the heart. Begin your incision at the posterior edge of one of the atria and continue downward to the *apex*, or tip of the heart. Then continue upward on the second side to the posterior border of the other atrium. Separate the two ventricles of the heart as in the photo, leaving the atria intact.

The upper part of the picture shows the inside of the ventral half of the sheep heart, the lower part shows the dorsal half.

Observe the thickness of the outer muscular wall (myocardium). Can you tell which is the right and which is the left side of the heart? The thickness of the wall will reveal this. To what parts of the body does the left ventricle pump blood? To what parts of the body does the right ventricle pump blood?

Note the *chordae tendineae*. At one end these tough, white tendinous "heart strings" are attached to the *atrioventricular* values (A-V valves). To what are they attached at the other end? Their function is to prevent the A-V valves from being turned inside out during the high pressure phase of ventricular systole. A backflow of blood into the atria is thus prevented.

Note the very musclular *interventricular septum*, simply labeled "septum" in the photo, a dividing wall between the right and left ventricles. It effectively separates oxygenated and deoxygenated blood in the ventricles and participates in the pumping action of the heart. A much thinner *interatrial septum* is found between the two atria.

Also note the large blood vessels entering and leaving the heart. Use wooden probes to trace these vessels. In which chamber of the heart does each originate or terminate? Can you identify these vessels?

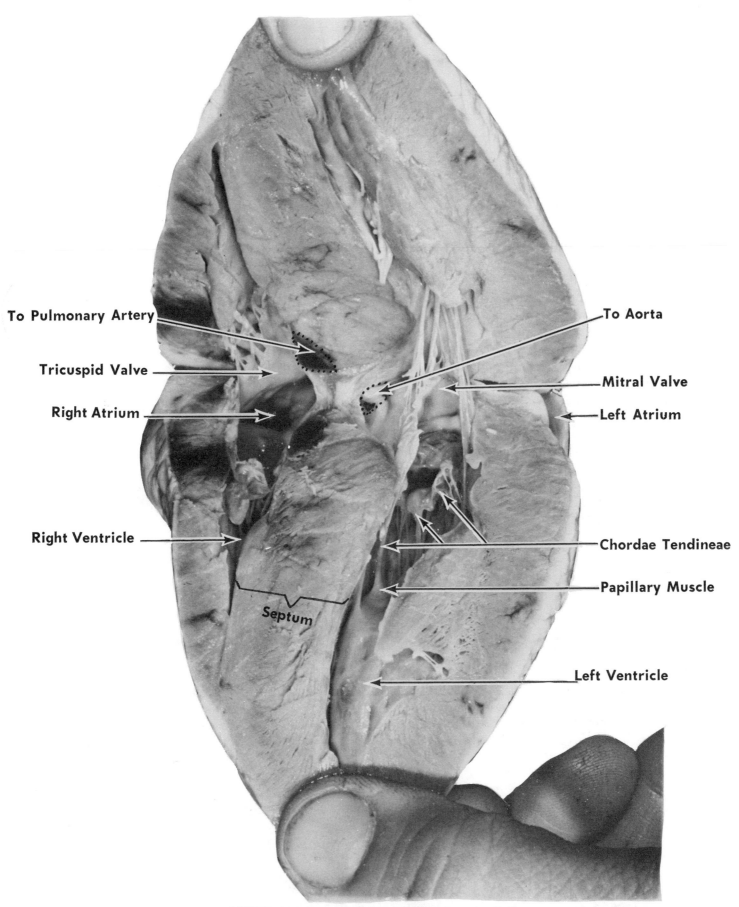

To Pulmonary Artery

Tricuspid Valve

Right Atrium

Right Ventricle

Septum

To Aorta

Mitral Valve

Left Atrium

Chordae Tendineae

Papillary Muscle

Left Ventricle

SHEEP HEART (Open), CORONAL PLANE

SHEEP HEART (OPEN), RIGHT SIDE

In order to expose the right side of the heart as in the photo, proceed as follows:

Open the *pulmonary artery* by cutting longitudinally along its lenght. Continue to cut through the *myocardium* of the *right ventricle* to the apex of the heart. Similarly, cut open the *superior vena cava*. Continue the cut through the wall of the *right atrium* until the two cuts cross. Now, spread apart the heart and observe.

Note:
- —The semilunar valve at the origin of the *pulmonary artery*.
- —The tree cusps, or flaps of tissue, of the *tricuspid valve* and the tough *chordae tendineae* holding the valve, in place.
- —The *papillary muscles* from which the chordae tendineae originate.
- —The *superior vena cava* entering the *right atrium*.
- —The inner rough textured walls of the right atrium.
- —The entrance for the *inferior vena cava*.
- —The opening of the *coronary sinus* bringing blood from the coronary veins and the flap-like valve at the entrance.

Compare the thickness of the walls of the atrium and ventricle. Relate these to their functions.

Find all of the structures labeled in the photo.

SHEEP HEART (OPEN), LEFT SIDE

You may use the same specimen of sheep heart as in the preparation seen in the last photo.

Turn the heart to the left side. Find the *aorta* and open it by cutting longitudinally along its length. Continue your incision posteriorly through the *myocardium* of the *left ventricle* to the apex of the heart. Then, cut open the left atrium along its lateral surface. Now, spread the heart apart and observe.

Note:
- —The *semilunar valves* at the base of the aorta.
- —The openings in the wall of the aorta where the *coronary arteries* and the *brachiocephalic artery* originate.
- —The two cusps, or flaps of tissue, of the *bicuspid (mitral) valve*, the *chordae tendineae* and the *papillary muscles*.
- —The inner surface of the *left atrium* and the opening of the *pulmonary veins*.
- —The *pulmonary artery* may also be seen.

Find all of the structures labeled in the photo.

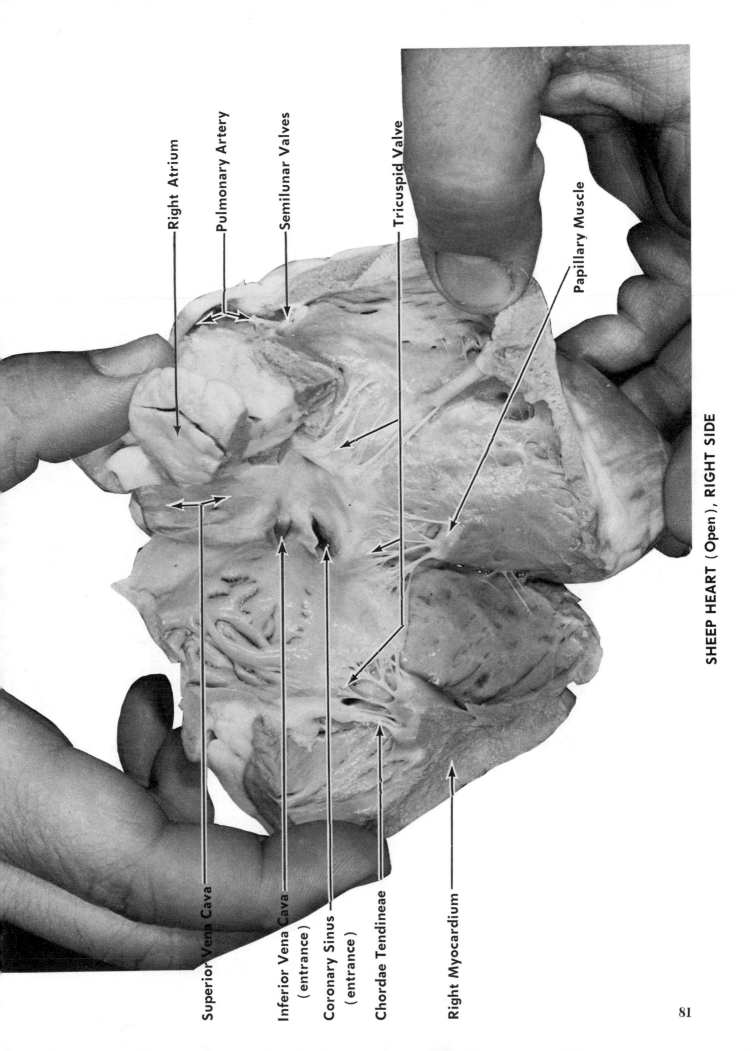

Right Atrium

Pulmonary Artery

Semilunar Valves

Tricuspid Valve

Papillary Muscle

Superior Vena Cava

Inferior Vena Cava
(entrance)

Coronary Sinus
(entrance)

Chordae Tendineae

Right Myocardium

SHEEP HEART (Open), RIGHT SIDE

81

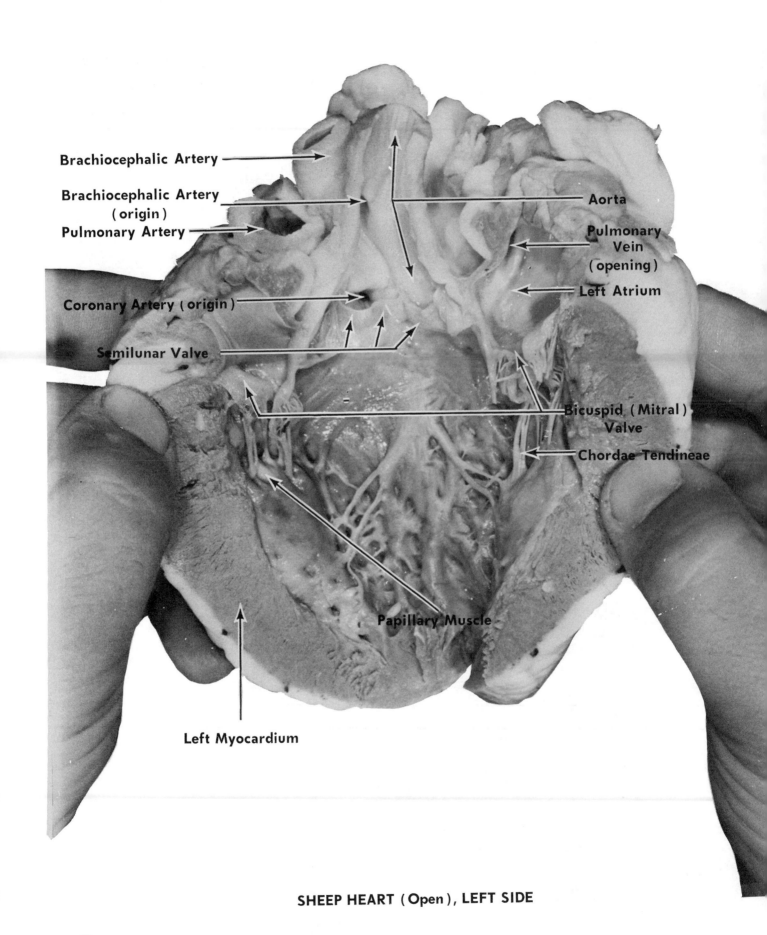

Brachiocephalic Artery

Brachiocephalic Artery (origin)

Pulmonary Artery

Aorta

Pulmonary Vein (opening)

Left Atrium

Coronary Artery (origin)

Semilunar Valve

Bicuspid (Mitral) Valve

Chordae Tendineae

Papillary Muscle

Left Myocardium

SHEEP HEART (Open), LEFT SIDE

THE HUMAN HEART

Aortic Arch

Pulmonary Artery

Superior Vena Cava

Openings to Coronary Artery

Pulmonary Semilunar Valve

Right Atrium

Tricuspid Valve

Opening to Coronary Sinus

Inferior Vena Cava

Chordae Tendineae

Papillary Muscles

Left Pulmonary Artery

Pulmonary Veins

Left Atrium

Bicuspid Valve

Aortic Semilunar Valve

Left Ventricle

Right Ventricle

Ventricular Septum

Myocardium

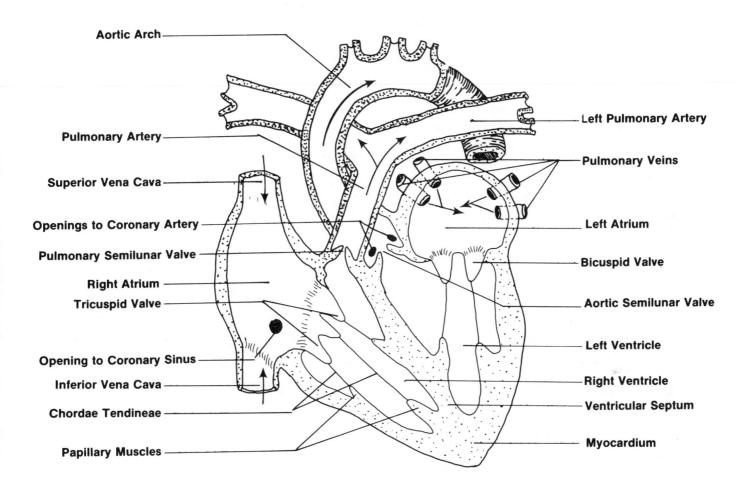

Note: Arrows Indicate The Path Of Blood Flow

SELF - QUIZ VI
THE HEART

1. Name the membrane sac which encloses the heart.
2. A large gland which extends into the neck region partially covers the heart. Name this gland.
3. Another name for the mitral valve is the . . . . valve.
4. Name the valves located at the origin of the aorta.
5. The first branches of the aorta, located near its base, lead to the . . . . and are known as the . . . . arteries.
6. Blood returning from the lower portion of the body enters the right atrium of the heart through the blood vessel known as the . . . .
7. The flaps of the atrioventricular valves are held in place by the tough fibrous . . . .
8. Oxygenated blood from the lungs enters the heart at the (a) right atrium, (b) left atrium, (c) right ventricle, (d) left ventricle.
9. The right and left ventricles are separated by the . . . .
10. Define each of the terms listed below.

ANSWERS

1. _____
2. _____
3. _____
4. _____
5. _____
6. _____
7. _____
8. _____
9. _____

10.   a.   papillary muscle _____

     b.   foramen ovale _____

     c.   endocardium _____

     d.   coronary thrombosis _____

     e.   fibrillation _____

     f.   mitral stenosis _____

     g.   heart murmur _____

     h.   rheumatic heart _____

     i.   chordae tendineae _____

     j.   tachycardia _____

Label all of the features indicated on the photograph.

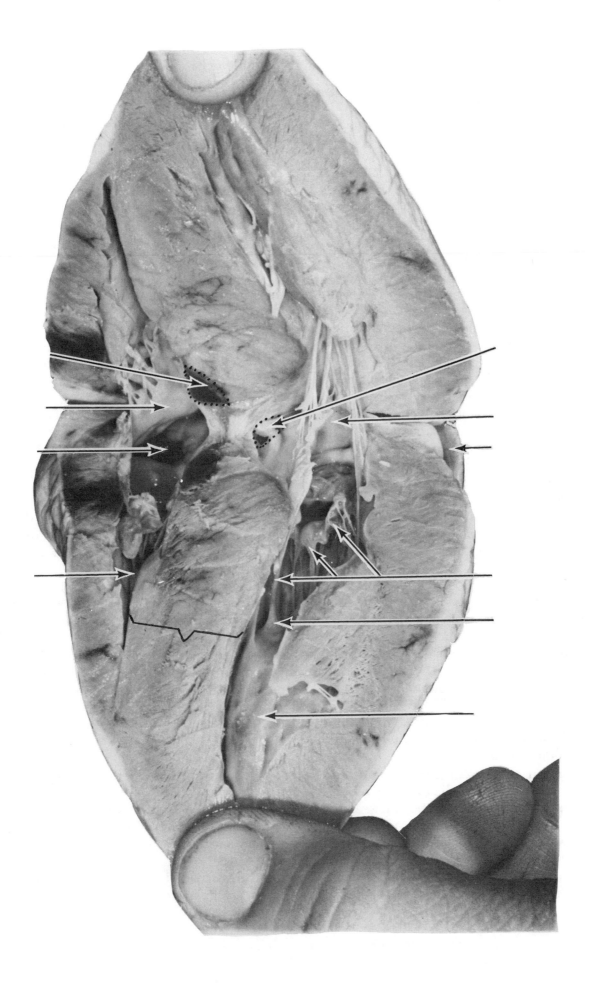

VEINS - ANTERIOR REGION

We begin our study of the major blood vessels with the veins which carry blood to the heart from the head, forelimbs, shoulder, and thoracic region.

The anterior veins lie close to the surface of the body and are therefore studied before the arteries which lie deep to the veins. In doubly injected specimens veins will be blue, arteries will be red. This is due to the latex dye with which they were injected. This permits you to see these vessels more clearly and to differentiate between arteries and veins.

You have already exposed the heart and some of the major veins in your study of the thoracic cavity. Extend your incision into the neck area. If you have not yet done so, expose the *trachea*, *larynx*, and *thyroid gland* as in the photo. Remove the *thymus gland* carefully from the ventral surface of the heart and lower trachea.

Your best dissection instrument at this point is the dissecting needle. Use it instead of a scalpel or scissors to expose and follow the path of a blood vessel. Scalpels and scissors, especially in the hands of the novice tend to destroy rather than preserve or expose. It is used for clearing a blood vessel of the connective tissue adhering to it, to tear into a thin muscle layer to follow a vessel, to separate one blood vessel from another, or from the nerves associated with it.

Proceed to expose the large veins surrounding the heart.

Anterior Vena Cava — Locate the trunk of this systemic vein above the heart. In the pig the term anterior vena cava is more correct and therefore preferred to superior vena cava. All anterior veins lead into the anterior vena cava. Are there any exceptions? Where does the anterior vena cava empty?

Posterior Vena Cava — This is the major vein returning blood from the lower extremities and from the abdominal area. Again, this designation in the pig is preferred to inferior vena cava. It can clearly be seen rising from the diaphragm, which it has penetrated, to enter the heart at the right atrium together with the anterior vena cava.

Innominate (Brachiocephalic) — Follow the anterior vena cava cranially from the heart. It divides into two equal branches, the "V" shaped *innominate veins* also known as the brachiocephalic veins. Each of these is formed by the union of the *subclavian vein* from the shoulder and the *external jugular vein* from the neck.

Jugular Veins — Two jugular veins lie along each side of the neck. The *external jugular vein* drains the head and neck, while the *internal jugular vein* drains the brain and spinal cord. They join to form the *innominate vein*.

Subclavian — This vein is a continuation of the *axillary vein* in the shoulder joint area. The axillary vein is a continuation of the brachial vein in the arm.

Internal Mammary (Sternal) **Vein** — On the ventral surface of the anterior vena cava about one inch above the heart the *internal mammary* or sternal vein arises. It results from the union of the paired internal mammary veins which drain the ventral thoracic body wall and the mammary glands of female pigs.

Subscapular — This vein drains the shoulder region and arises from the underside of the scapula. It enters the *subclavian vein* close to the base of the external jugular vein.

Cephalic — This vein lies upon the lateral surface of the foreleg and was seen when the pig was skinned. It is a long vein which continues up to the shoulder ara and inters the external jugular vein at its base. The *cephalic vein* and the deeper vein of the arm, the *brachial vein*, are joined in the inner elbow by the *median cubital vein.*

Long Thoracic — This vein arises along the inner surface of the pectoralis minor muscle then passes to the latissimus dorsi. It joins the *axillary vein.*

Thoracodorsal — This vein arises from under the scapula and similarly joins the *axillary vein.*

Hemiazygous — Push the left lung and the heart medially. On the dorsal body wall, to the left of the midline, and running parallel to the aorta, observe a vein which enters the anterior vena cava near its entrance into the right ventricle of the heart. This unpaired vessel is the *hemiazygous vein.* It drains the dorsal body wall and the dorsal musculature. Note that the *intercostal veins* between the ribs empty into the hemiazygous vein. Dissect the pairs of intercostal veins to observe this relationship. In man, the hemiazygous vein is replaced with the *azygous vein* which lies on the right side.

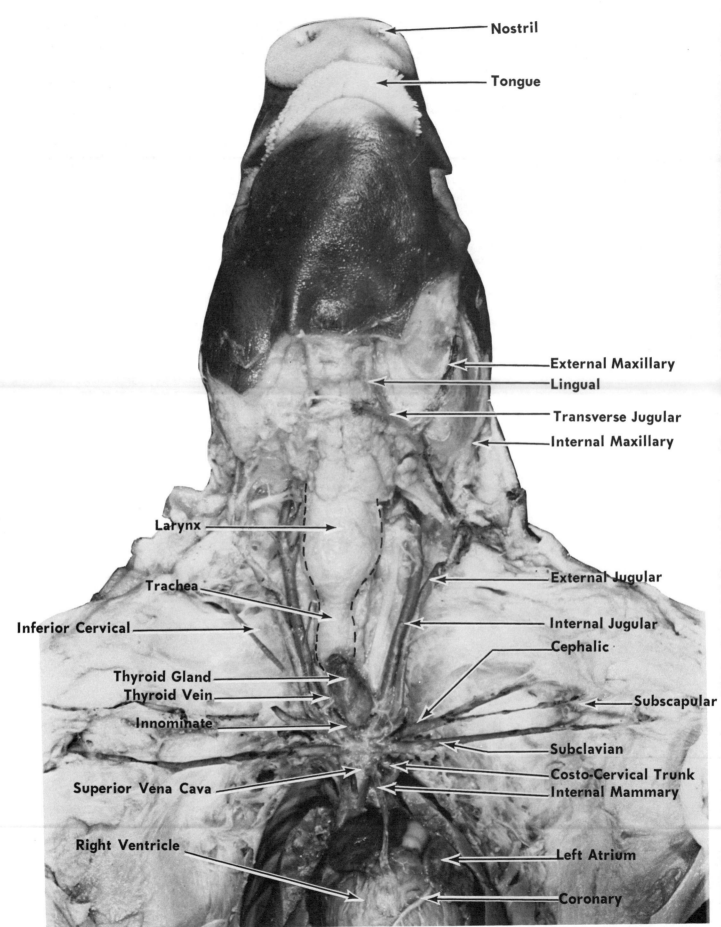

Nostril

Tongue

External Maxillary

Lingual

Transverse Jugular

Internal Maxillary

Larynx

External Jugular

Trachea

Inferior Cervical

Internal Jugular

Cephalic

Thyroid Gland

Thyroid Vein

Subscapular

Innominate

Subclavian

Costo-Cervical Trunk

Superior Vena Cava

Internal Mammary

Right Ventricle

Left Atrium

Coronary

ANTERIOR VEINS

88

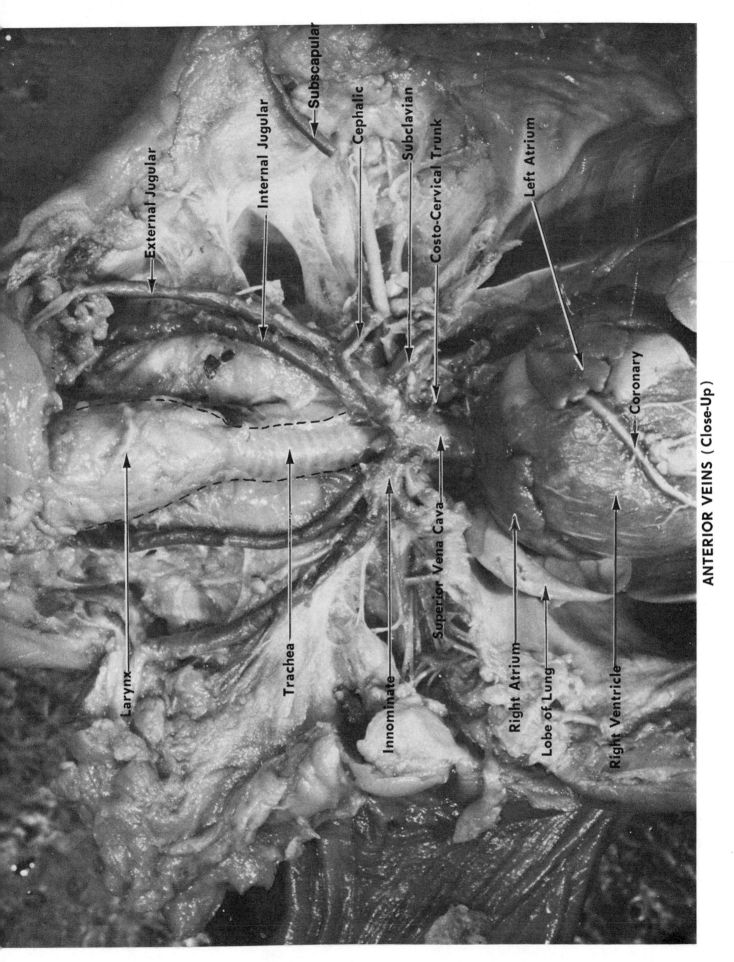

External Jugular

Internal Jugular

Subscapular

Cephalic

Subclavian

Costo-Cervical Trunk

Left Atrium

Larynx

Trachea

Innominate

Superior Vena Cava

Coronary

Right Atrium

Lobe of Lung

Right Ventricle

ANTERIOR VEINS (Close-Up)

VEINS - POSTERIOR REGION

The systemic veins below the heart empty into the *posterior vena cava*. The upper portion of this enlarged blood vessel was observed when we studied the thoracic veins. It passes through the diaphragm and lies along the mid-dorsal body wall of the thorax. It then enters the right atrium of the heart near the entrance of the anterior vena cava.

Posterior Vena Cava — This large blood vessel may be observed by displacing the abdominal viscera toward the left. It lies along the mid-dorsal abdominal body wall. Its major branches will be described in descending order, in a caudal direction.

Phrenic — Within the diaphragm the posterior vena cava receives these veins.

Hepatic — Below the diaphragm region the posterior vena cava passes through the substance of the liver within which it receives the veins which drain the liver, the hepatic veins.

Ductus Venosus — Within the body of liver the inferior vena cava then receives the ductus venosus. This vessel is a continuation of the *umbilical vein* which you observed earlier in the dissection. As you recall, it was necessary to cut this vein in order to lift up the ventral body wall to view the abdomen. The oxygen rich blood from the mother passes to the fetus through the umbilical vein within the umbilical cord. The umbilical vein leads from the cord to the liver where it forms the ductus venosus. In order to see the relationship between these veins it is necessary to remove some of the substance of the liver. Use your forceps or dissecting needles. You will discover a rich network of blood vessels running through the liver. The ductus venosus drains into the posterior vena cava posterior to the hepatic veins.

Renal — You have already observed the large kidneys lying laterally in the anterior portion of the abdominal cavity. Find the renal vein exiting from the medial surface of each kidney and trace it to the posterior vena cava. In some specimens two renal veins drain each kidney.

Adrenal — The adrenal gland is an alongated oval structure lying medial to the anterior portion of the kidney. It is drained by the adrenal vein which empties the renal vein. In some specimens it enters the posterior vena cava separately.

Genital *(Utero-Ovarian and Internal Spermatic)* — The genital vein is known by different names in males and females. These are very narrow and difficult to locate and are most easily traced from the ovaries or the testes. The right genital vein enters the posterior vena cava next to the posterior end of the right kidney, while the left genital vein generally enters the renal vein.

Lumbar — Posterior to the kidneys seven pairs of veins drain the deep dorsal and lateral abdominal walls. The upper five to six pairs empty into the posterior vena cava, while the last pairs empty into the *common iliac veins*.

Common Iliac — The venous blood from the pelvic region and from the legs moves anteriorly toward the heart. The right and left common iliac veins serve as receiving vessels for all of this blood. They converge to form the *posterior vena cava*, forming the shape of an inverted "Y". All veins below this point are branches of the common iliac veins.

External and Internal Iliac — The common iliac is a relatively short segment of blood vessel. Near its origin it may be seen as formed by two branches. The main, larger and thicker branch extends laterally. This is the *external iliac*. It extends out toward the legs. A shorter and narrower vessel, the *internal iliac (hypogastric) vein* extends posteriorly on either side of the midline. It drains the rectum, urinary bladder, and the genital organs.

> **Circumflex Iliac** — This vein is a lateral branch of the *external iliac vein*. It drains the pelvic muscles.
>
> **Femoral** — In the thigh the *external iliac vein* is known as the *femoral vein*. It receives several branches.
>
> > **Deep Femoral** — This vein drains the deeper muscles of the thigh.
> >
> > **Saphenous** — This superficial vein may be seen lying on the medial surface of the thigh and lower leg. Above the knee it joins the *femoral vein*.
> >
> > **Popliteal** — Posterior to the knee the femoral vein is joined by a deeper vein from the lower thigh and shank, the popliteal vein.

Middle Sacral — This, often paired vein, runs along the mid dorsal pelvic wall draining the musculature. It empties into the *common iliac veins*.

Caudal — This vein, extending into the tail, receives blood from the most posterior structures and the musculature of the tail. Anteriorly it joins the middle sacral vein.

Hepatic Portal System — The veins from most of the abdominal viscera do not join the posterior vena cava directly. Instead, blood from the stomach, pancreas, spleen, small intestine, and large intestine drain into a larger vein known as the *hepatic portal vein* which enters the liver and there breaks up into capillaries called *sinusoids*. The hepatic portal vein is rich in digested nutrients. Before these enter the general circulation the liver transforms them according to the needs of the body. Glucose may be stored as glycogen, amino acids deaminated, or fatty acids may be converted to carbohydrates.

In order to expose the hepatic portal vessels move the stomach, spleen, pancreas, small, and large intestines to the left. Frequently they are not injected with blue latex even in injected specimens, therefore, only the largest vessels can be readily identified.

> **Hepatic Portal Vein** — This is the main vein of the hepatic portal system. All of the tributary vessels of the abdominal viscera merge with this single large vein. It enters the liver where it breaks up into smaller vessels then into capillaries. These are drained by the hepatic veins and posterior vena cava as previously described (see p. 90). Some of the major veins joining the hepatic portal are:
>
> > **Anterior Mesenteric** — This vein is a union of branches from the many coils of the small intestines.
> >
> > **Gastroduodenal** — This vein drains the pyloric region of the stomach and the duodenum before it joins the hepatic portal vein.
> >
> > **Gastrosplenic** — As its name indicates, this vein originates at the stomach and spleen.

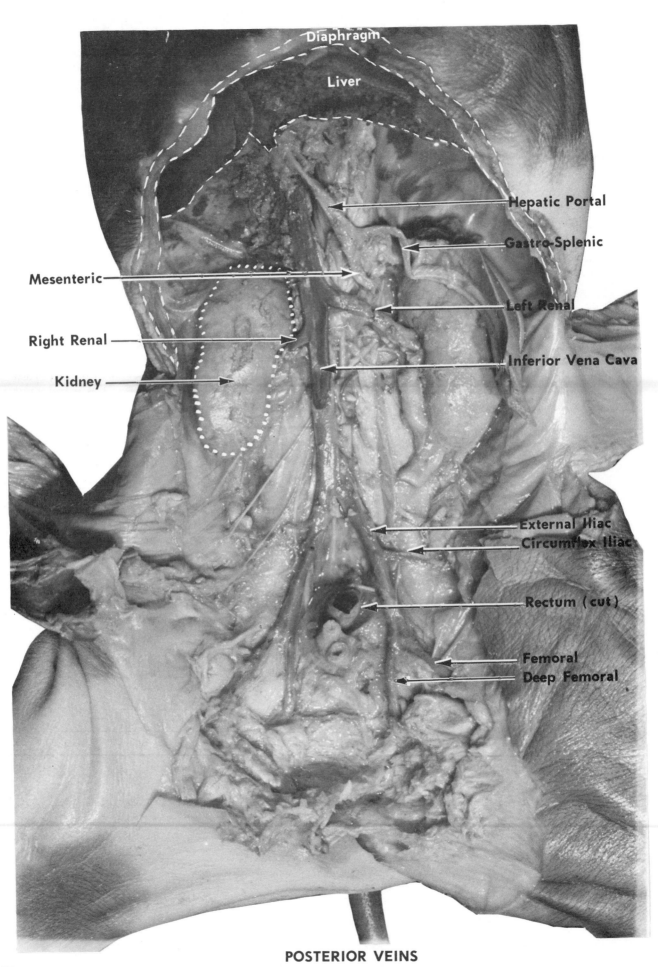

Diaphragm

Liver

Hepatic Portal

Gastro-Splenic

Mesenteric

Left Renal

Right Renal

Inferior Vena Cava

Kidney

External Iliac

Circumflex Iliac

Rectum (cut)

Femoral

Deep Femoral

POSTERIOR VEINS

ARTERIES - ANTERIOR REGION

Remove the major veins surrounding the heart. This will enable you to expose the arteries. Use your dissecting needle to clear arteries of connective tissue, to separate them, and to follow them.

Pulmonary Artery — On the ventral surface of the heart passing dorsally and to the left is the large *pulmonary artery*. It originates in the right ventricle. Try tracing its two branches to the lungs. The pulmonary is the only artery injected with blue latex to indicate that it carries deoxygenated blood. All other arteries will appear red or pink because they have been injected with red latex to indicate that the blood transported in them is oxygenated.

Aorta — Locate the *aorta*, the largest systemic artery of the body. It leaves the left ventricle, curves to the left, dorsal to the pulmonary artery, and continues dorsally in a posterior direction along the left side of the vertebral column. The proximal curved portion of aorta is called the *aortic arch*, while the next segment of the aorta within the thorax is known as the *thoracic aorta*.

Ductus Arteriosus — In the fetal pig, as well as in all mammalian fetuses, the pulmonary artery is joined directly to the aortic arch by means of a short vessel, the *ductus arteriosus*. It serves as a bypass to shunt the blood from the lungs toward the systemic circulation. This connecting link is about ¼ inch long in the older fetal pigs and is clearly seen in the photo on p. 95. The ductus anteriosus persists as a connecting link till birth. It then shuts tightly, separating the pulmonary artery from the aorta and the two major circulatory pathways, the *pulmonary* from the *systemic*. It persists in the adult as a narrow tendinous band.

Coronary — Near its origin within the left ventricle of the heart, the aorta gives off its first two branches, the *right* and *left coronary arteries*. Branches of these vessels may be seen upon the surface of the heart. These supply blood to the heart muscle (myocardium) directly.

The aortic arch gives rise to arteries that supply the neck, head, shoulders and forelimbs. Whereas in man *three* arterial trunks arise from the aortic arch, in the pig there are only *two*. These are the:

1. **Brachiocephalic (or Innominate)** — This major artery is the first of the two trunks branching from the aortic arch. As its name indicates, it supplies blood to the forelimb and the head. At the level of the second rib it divides into the:

> **Right Subclavian** — This artery leaves the thorax, gives off major branches to the musculature of the shoulder and thorax then continues as the:

>> **Axillary** — This artery is a continuation of the right subclavian in the region of the armpit and shoulder.

>> **Brachial** — This is the continuation of the axillary artery in the upper forelimb.

>> **Radial** and **Ulnar** — These are branches of the brachial artery within the lower forelimb.

> **Bi-Carotid Trunk** — This artery is the second branch of the brachiocephalic. It is only a short common pathway for blood to the head (see photo, p. 95). It soon divides to form the:

>> **Right and Left Common Carotids** — These two arteries arise from the bi-carotid trunk, pass cranially, parallel to one another on eigher side of the trachea. Near the head each divides into an *internal* and *external carotid artery*.

2. **Left Subclavian** — The second branch of the aortic arch is the left subclavian artery. It gives off branches to the musculature of the shoulder, then proceeds into the left forelimb as the *left axillary*, the *left brachial*, the *left radial*, and the *left ulnar* arteries.

The *subclavian artery* gives off major branches to the musculature of the thorax and shoulder. Starting from its proximal end these include the:

Costo-Cervical Trunk — This is the first vessel to come off the subclavian artery. It originates from the dorsal surface of the subclavian at the level of the first rib. Some of its branches are the:

Vertebral — This is the most anterior branch of the *costo-cervical trunk*. The artery passes through the transverse foramina (channels) of the cervical vertebrae cranially toward the brain.

Deep Cervical — This artery sends branches to the anterior intercostal muscles.

Dorsal — This small artery is the third branch of the **costo-cervical trunk**. It supplies blood to lower neck and anterior part of the back regions.

Thyrocervical (or Inferior Cervical) — This comparatively large artery arises from the dorsal surface of the subclavian near the first rib. It supplies the thyroid gland, pectoral muscles, and parotid gland.

External Thoracic (Lateral Thoracic) — This arterial branch of the subclavian supplies blood to both the superficial and deep thoracic muscles.

Internal Mammary (Internal Thoracic) — This artery also originates from the subclavian opposite to the origin of the thyrocervical. It passes posteriorly to supply the pectoral muscles, the skin, the thymus gland, and the mammary gland. It is usually cut when exposing the thoracic cavity.

Subscapular — This artery originates just beyond the subclavian, in the *axillary artery*. It passes anteriorly to the shoulder region where it supplies the subscapular region with blood.

Thoraco-Dorsal — This artery is a branch of axillary arteria. It originates opposite to the subscapular and passes posteriorly. In some specimens it is a branch of the subscapular artery.

We have traced the *common carotid* artery to the head. Just below the jaw it divides to form the *internal carotid artery* and the *external carotid artery*.

Internal Carotid — This artery runs anteriorly into the skull to supply the brain. Near its origin it gives off the *occipital artery* which passes dorsally to supply the occipital region.

External Carotid — This artery supplies the remainder of the head with blood. Some of its branches include the:

Lingual — This artery supplies the tongue.

External and Internal Maxillary — to the lower and upper jaws, the lips, and mouth

Posterior Auricular — to the ear

Superficial Temporal — found near the temple

Return to the *aortic arch* Trace its path to the left as it descends mid-dorsally within the thorax. It will be necessary to displace the left lung medially in order to observe this segment of the aorta. It is here known as the *thoracic dorsal aorta*. Upon closer examination, you will find paired arterial branches originating along the length of the thoracic aorta. These pass into the musculature between the ribs, and are known as the *intercostal arteries*. While most of them originate in this way directly from the aorta, the upper few originate from the *costo-cervical trunk*.

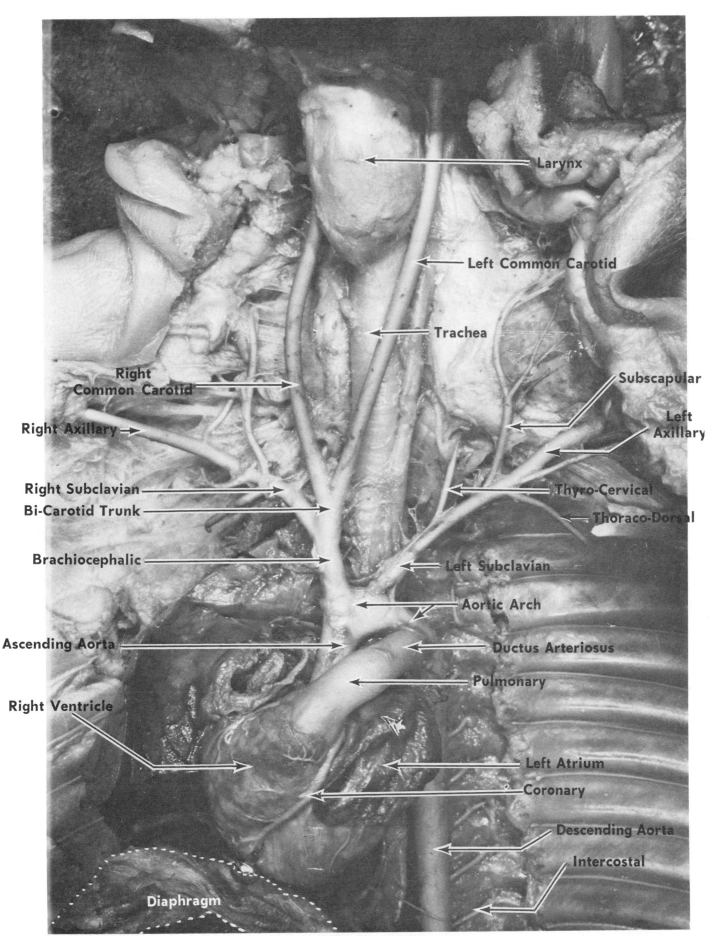

Larynx

Left Common Carotid

Trachea

Right Common Carotid

Subscapular

Right Axillary

Left Axillary

Right Subclavian

Thyro-Cervical

Bi-Carotid Trunk

Thoraco-Dorsal

Brachiocephalic

Left Subclavian

Aortic Arch

Ascending Aorta

Ductus Arteriosus

Pulmonary

Right Ventricle

Left Atrium

Coronary

Descending Aorta

Intercostal

Diaphragm

ANTERIOR ARTERIES

ARTERIES - POSTERIOR REGION

Follow the *aortic arch* dorsally and caudally through the thorax. Push the left lung toward the right and observe the *thoracic aorta* along the dorsal body wall to the left of the vertebral column.

Intercostal Arteries — Between each two ribs note the *intercostal arteries.* These were already described earlier (see p. 94). They are given off by the *thoracic aorta* in pairs, to the right and left sides, to supply the intercostal muscles.

Other branches of the *thoracic aorta* above the diaphragm include the:

Bronchial — These arteries supply the lung tissues with oxygenated blood.

Esophageal — This artery supplies the esophagus.

Phrenic — This artery supplies blood to the diaphragm.

Trace the dorsal aorta as it continues caudally and passes through the diaphragm into the abdominal cavity. It is here known as the *abdominal aorta.* It gives rise to a number of arteries supplying the organs of the abdominal region. These include the:

Celiac — Immediately posterior to the diaphragm the abdominal aorta gives off ventrally a short unpaired blood vessel, the *celiac artery,* which branches extensively to supply blood to the organs of the upper abdomen. It divides into two main branches:

    1. **Gastro-Splenic** — This artery supplies blood to the stomach and to the spleen.

        Gastric — This portion of the gastro-splenic artery passes to the underside of the stomach.

        Splenic — The second portion of the gastro-splenic passes medially along the length of the spleen to supply blood to that organ.

    2. **Gastro-Hepatic** — This branch supplies blood to other portions of the stomach and to the liver. It passes through the *pancreas* to supply blood to that organ. Then it divides to form the:

        Hepatic — This artery supplies the liver.

        Gastro-Duodenal – This artery supplies the duodenum and stomach.

Anterior Mesentric — This is the next arterial branch of the abdominal aorta below the celiac artery. It supplies the small intestine and the anterior portion of the large intestine.

Phrenico-Abdominal — This artery comes off the aorta near the anterior end of the kidney. It supplies the diaphragm and the lateral abdominal body wall with blood.

Renal — Each kidney is supplied with blood by the renal artery.

Adrenal — The adrenal gland is supplied by the adrenal arteries which may originate directly from the abdominal aorta or from the renal artery.

Genital — These arteries supply the male and female glands with blood. They are called:

    Spermatic — These arteries pass posteriorly to become part of the *spermatic cord* to the testes within the scrotal sac in males.

    Utero-Ovarian — The corresponding artery in females leads to the *uterus* and *ovaries.*

Posterior Mesentric — This unpaired vessel originates from the aorta below the genital artery. It divides to supply blood to the anterior as well as the posterior portions of the large intestines.

Lumbar — Follow the abdominal aorta in the mid-dorsal region of the abdomen. Note six pairs of arteries passing laterally to the dorsal muscle wall. Those are the *lumbar arteries.*

At the level of the hip the aorta divides into two major vessels, one to each hind leg.

External Iliac — This artery gives off many branches to the organs of the pelvis, the genital and structures and to the hind legs.

Note: Unlike the venous system, there is no common iliac artery. The external and internal iliac arteries arise as separate branches of the aorta.

Circumflex Iliac — This lateral branch of the external iliac artery supplies blood to the abdominal wall.

Internal Iliac — As already indicated, this artery arises from the aorta at the same level as the external iliac. It is, however, more medially located, thus serving the organs of the pelvis.

Umbilical — The narrow internal iliac arteries soon enlarge as the *umbilical arteries*. These major prominent features of the fetal circulation carry deoxygenated blood from the developing fetus to its mother. The arteries lie on either side of the elongated *urinary bladder* before exiting the body. They were already observed when the ventral abdominal wall was removed. After birth these arteries atrophy. Beyond the umbilical arteries, the internal iliacs continue posteriorly within the pelvis.

External Iliac (continued) — Return to the external iliac artery.

Femoral — The segment of the extrnal iliac within the thigh is known as the *femoral artery*. It supplies the hind leg with blood.

Deep Femoral — This artery is a branch of the femoral artery. It supplies the medial upper region of the hind leg.

Saphenous — Just above the knee joint the femoral artery gives off the saphenous artery which passes medially down the leg.

Popliteal — This segment of the femoral artery originates at the back of the knee cap and passes to the lower portion of the hind leg.

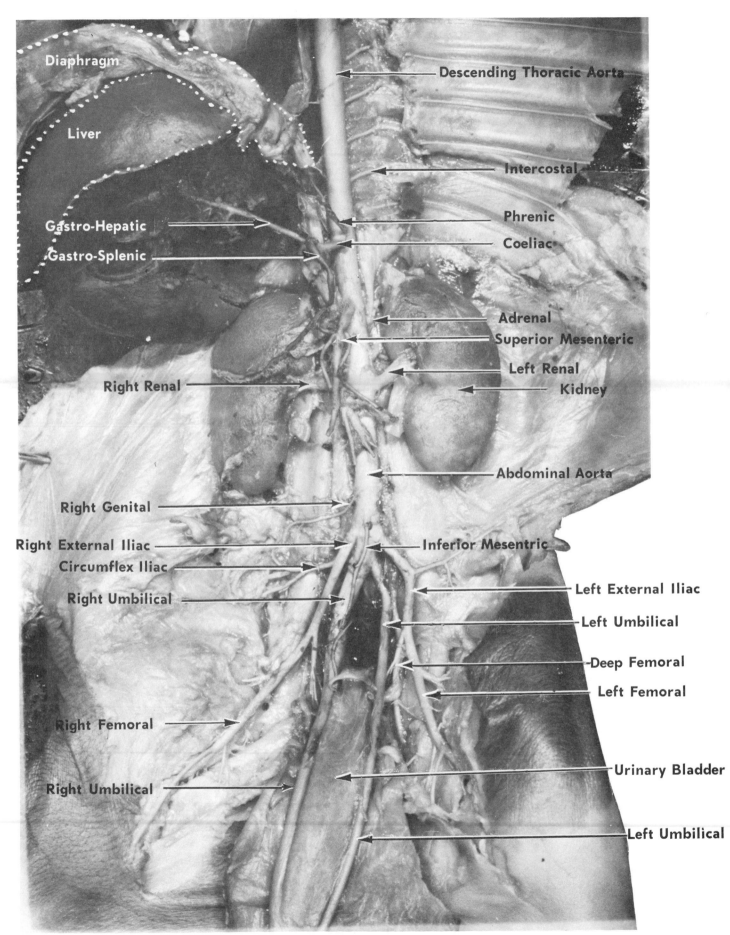

Diaphragm

Liver

Gastro-Hepatic

Gastro-Splenic

Right Renal

Right Genital

Right External Iliac

Circumflex Iliac

Right Umbilical

Right Femoral

Right Umbilical

Descending Thoracic Aorta

Intercostal

Phrenic

Coeliac

Adrenal

Superior Mesenteric

Left Renal

Kidney

Abdominal Aorta

Inferior Mesentric

Left External Iliac

Left Umbilical

Deep Femoral

Left Femoral

Urinary Bladder

Left Umbilical

POSTERIOR ARTERIES

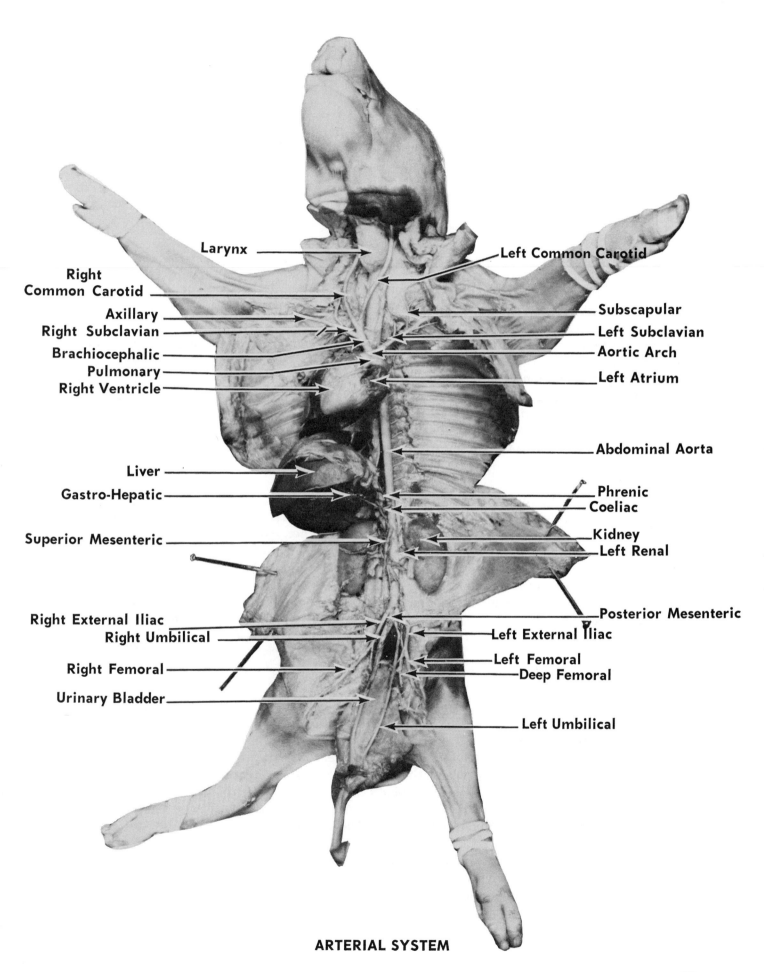

Larynx

Right
Common Carotid

Axillary

Right Subclavian

Brachiocephalic

Pulmonary

Right Ventricle

Liver

Gastro-Hepatic

Superior Mesenteric

Right External Iliac

Right Umbilical

Right Femoral

Urinary Bladder

Left Common Carotid

Subscapular

Left Subclavian

Aortic Arch

Left Atrium

Abdominal Aorta

Phrenic

Coeliac

Kidney

Left Renal

Posterior Mesenteric

Left External Iliac

Left Femoral

Deep Femoral

Left Umbilical

ARTERIAL SYSTEM

THE FETAL CIRCULATION

No other anatomical system is as unique in the mammalian fetus as the *circulatory system*. The muscles, glands, and most organs are merely smaller versions of the adult forms. The bones too are well formed although some have not ossified and are still partially cartilaginous.

In the circulatory system, however, we find structures of the fetus that are to be completely eliminated or drastically modified in the adult.

Overall, while circulation in the fetus of mammals is similar to that in adults, it differs in two principal features:

1. Most of the pulmonary circulation is made to bypass the lungs.

2. The fetal blood is sent to the uterus for oxygenation, removal of wastes, and the addition of nutrients and other materials (hormones, antibodies, etc.) which are needed by the developing fetus. This is necessitated by the the fact that the fetal lungs, digestive tract and kidneys have not yet assumed their adult function. After the exchanges within the placenta have been made, blood returns to the fetus.

Fetal Circulation To and From the Placenta

Most of the fetal blood flows into the *umbilical arteries* which pass out of the abdomen through the *umbilical cord* to the *placenta* within the *uterus* of the mother. After the exchange of gases, wastes, and nutrients the blood returns through the *umbilical vein* within the umbilical cord, enters the abdominal cavity and moves anteriorly towards the liver. Within the liver most of the blood flows through the *ductus venosus* which joins the *posterior vena cava*. Some flows into the *hepatic portal* vein to enter the liver and tissues, then leaves by way of the *hepatic veins* into the posterior vena cava. Thus far, there has been some mixing of oxygenated blood from the umbilical vein and deoxygenated blood of the posterior vena cava and the hepatic portal vein.

Fetal Circulation Within and Outside the Heart

Most of the blood returning to the heart by way of the posterior vena cava into the *right atrium* passes directly into the *left atrium* through a one-way opening in the *septum* between the atria known as the *foramen ovale*. This effectively bypasses the pulmonary circulation and the lungs.

Most of the blood from the *anterior vena cava* entering the right atrium does pass into the right ventricle. It leaves by way of the *pulmonary* artery. Very little of this blood gets to the lung due to another bypass. This time a short vessel, the *ductus arteriosus*, a direct connecting link between the pulmonary artery and the *aorta*, leads the blood into the systemic circulation of the aorta.

Thus, effectively, the blood has bypassed the lungs twice, once by way of the foramen ovale within the heart, then by the ductus arteriosus outside the heart.

Circulatory System Changes That Take Place After Birth

1. **Ductus Arteriosus** — This vessel becomes a tough connective tissue, the *ligamentum arteriosum*. The blood flow in the pulmonary artery is thus completely separated from that in the aorta.

2. **Foraman Ovale** — Due to the increase in blood pressure within the left side of the heart after birth, the one-way valve controlling the foraman ovale shuts. The tissues unite with the interatrial septum, leaving only a slight depression, to completely separate the right from the left atrium.

3. **Umbilical Arteries** — These become modified so that the segments that lay along the bladder become the *round ligaments of the bladder*. The remainder are modified as functional arteries in the adult.

4. **Umbilical Vein** — This structure becomes the *round ligament of the liver*. It serves to suspend the liver from the ventral body wall.

5. **Ductus Venosus** — This vessel within the liver closes, becomes fibrous, and is known as the *ligamentum venosum* in the adult.

6. **Umbilical Cord** — This structure dries and falls off several days after birth. The "scar" remaining throughout life is the *umbilicus* (navel).

FETAL CIRCULATION

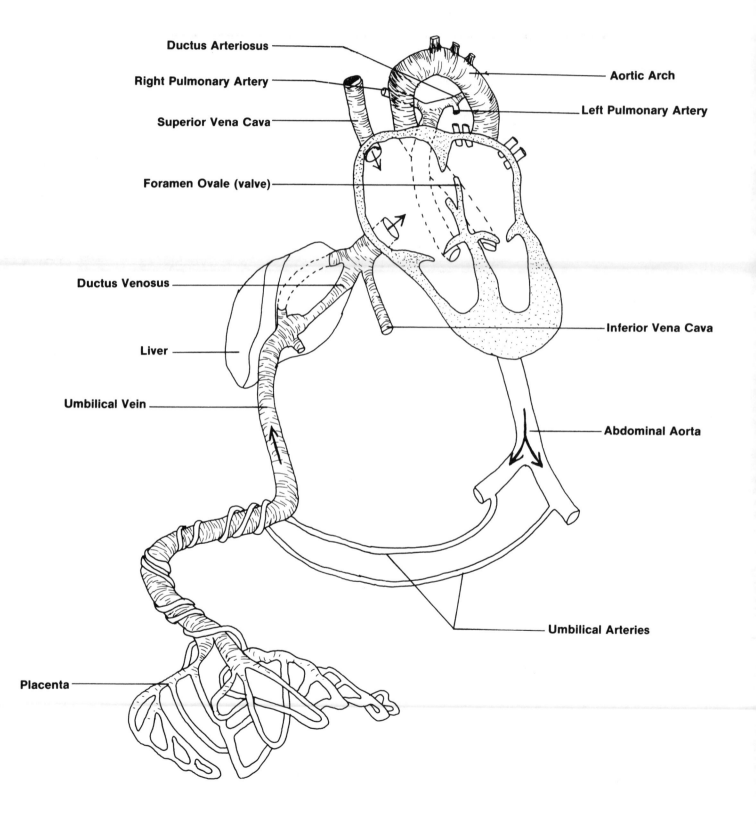

Ductus Arteriosus

Right Pulmonary Artery

Superior Vena Cava

Foramen Ovale (valve)

Ductus Venosus

Liver

Umbilical Vein

Placenta

Aortic Arch

Left Pulmonary Artery

Inferior Vena Cava

Abdominal Aorta

Umbilical Arteries

SELF - QUIZ VII
VEINS AND ARTERIES

1. Name the large veins and arteries that carry blood from and to the kidneys.
2. The jugular veins drain the head of blood. Name the large arteries lying alongside the jugulars that carry blood to the head.
3. The ductus arteriosus is a small blood vessel found in the fetal pig. It connects the pulmonary artery to the _____.
4. Name the blood vessels supplying the heart muscle with blood.
5. Name the branch of the aorta that extends to the shoulder and arm.
6. Besides the pulmonary vein, name another vein of the fetal pig that carries oxygenated blood.
7. The esophagus and two major blood vessels pass through the diaphragm. Name the blood vessels.
8. Veins from the abdominal viscera join to form a much larger vein which enters the liver. Name this large vein.
9. The abdominal aorta divides into two branches before entering the legs. Name these branches.
10. For each of the veins named below, tell from what parts of the body they carry blood.

ANSWERS

1. vein _____

   artery _____

2. _____

3. _____

4. _____

5. _____

6. _____

7. _____

   and _____

8. _____

9. _____

   and _____

10. a. innominate _____

    b. femoral _____

    c. intercostals _____

    d. saphenous _____

    e. gastro-splenic _____

    f. mesenteric _____

    g. brachial _____

    h. cephalic _____

    i. radial _____

    j. phrenic _____

Label all of the features indicated on the photograph.

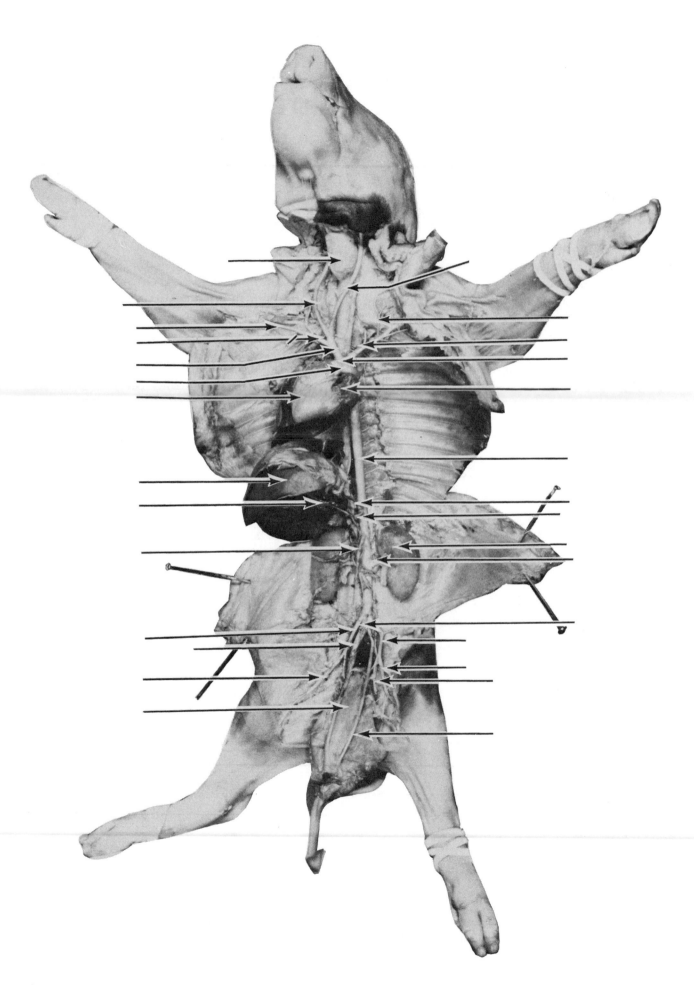

THE UROGENITAL SYSTEM - FEMALE

The *urinary* and *genital* systems have distinct and unique functions. The first serves to remove nitrogenous and other wastes and to maintain the body's water balance, while the other functions in the reproduction of the species. However, due to the similarities of their developmental origins and the sharing of common structures, they are usually considered as a single system, the *urogenital system.*

We will first study the urinary system, which is similar in males and females. We will then proceed to study the reproductive systems of the two sexes.

You are responsible for learning the reproductive systems of both male and female pigs. After studying your specimen, examine the reproductive system of a specimen of the opposite sex.

Remove the liver, spleen, stomach, and intestines. Leave the last two inches of the large intestine intact.

Urinary System
Kidneys — They are large bean-shaped structures on either side of the vertebral column at the level of the third to fifth lumbar vertebrae. Although they bulge into the abdominal cavity, they lie beneath the peritoneum, or *retroperitoneally,* often surrounded by fat. The *adrenal glands* are narrow band-like structures lying median to the anterior region of the kidneys.

Clear the kidneys to expose the *renal arteries, renal veins,* and the *ureters.* Some of the parts of the kidney are the:

Hilus — This is a central depression in the medial surface of the kidney. The ureters exit the kidney at the hilus.

In order to observe the following structures it is necessary to cut one of the kidneys in frontal section as in the diagram below.

Renal Sinus — This is a central cavity which contains fat, branches of the renal vessels and the *renal pelvis.* The pelvis is the funnel-shaped expanded portion of the ureter within the renal sinus.

Renal Cortex — This is the outer layer of kidney tissue.

Renal Medulla — This is the more central portion of the kidney, beneath the cortex.

Renal Papilla — This is a cone-shaped projection of the medulla enclosed by the pelvis. In the pig's kidney there is only one papilla, in man there are many.

Trace the ureter from the hilus to the *urinary bladder.* Do this on both sides. Do not injure the reproductive structures. Lift the urinary bladder and find the *urethra,* which transports urine from the bladder.

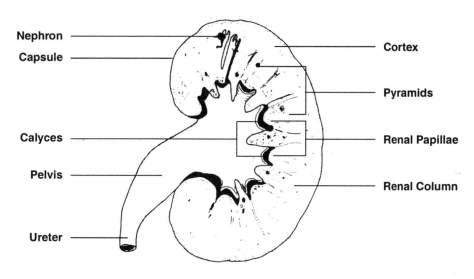

HUMAN KIDNEY (Frontal Section)

To this point only the urinary structures have been examined. They are alike in males and females.

If your specimen is a female, continue the dissection as directed here. If your specimen is male, continue in the next section. However, whether your specimen is male or female, you are responsible for knowing the reproductive structures of each. Therefore, work closely with a student whose pig is of the opposite sex of your specimen.

Genital System

Ovaries — These are the female gonads. They are paired, small bean-shaped structures located posterior to the kidneys. The *oviducts*, or fallopian tubes, and extremely narrow, usually located dorsal to the anterior portion of the ovary. Use a hand lens to observe them more closely. Also observe the expanded ends of the openings, the *ostium*, fringed by small finger-like projections, termed *fimbriae*. These guide the ova into the ostium.

Uterine Horns (Or Cornua) — Trace the oviducts to the dorsal surface of the ovary where they join the much wider *uterine horns*. Trace these caudally to where they join to form the body of the *uterus*, which lies dorsal to the urinary bladder and urethra.

In pigs and other mammals, the fetus does not develop in the body of the uterus, as in man, but in the horns extending from the uterus. This permits the development of more fetuses at one time and the birth of a litter. In humans, development of the fetus in the body of the uterus makes multiple births a rarity.

Membranes — The ovaries are suspended from the dorsal body wall by a peritoneal membrane called the *mesovarium*. An *ovarian ligament* connects the ovaries to the uterine horns. Each horn is supported by a peritoneal fold the *mesometrium*. These three membranous suspensions are part of the *broad ligament*. This ligament extends into the pelvic area serving to hold the body of the uterus and vagina to the body wall. Another support, the *round ligament*, extends from the dorsal body wall to the middle of each uterine horn.

In order to continue the dissection it is necessary to cut through the *pubic symphasis* and spread the pelvic bones apart. This will expose the rest of the genital organs lying below the pelvic bones.

Uterus and Vagina — You are now ready to expose the entire urethra and the body of the *uterus*. The *vagina* is a continuation of the uterus and lies dorsal to the urethra. The *cervix* is a constricted area between the body of the uterus and the vagina.

Urogenital Sinus — Separate the urethra from the vagina. Posteriorly the vagina and urethra unite to form a common passageway which opens to the exterior, the *urogenital sinus*, or *vestibule*. In human females the vagina and urethra are separate throughout their lengths and the vestibule is a much reduced area, a part of the external genitalia.

External Genitalia — Follow the urogenital sinus caudally to its opening on the outside of the body. Use your scissors to make a ventral longitudinal cut along the urogenital sinus in an anterior direction from the external genitalia, through the vagina beyond its union with the urethra, as in the accompanying photo. Find the *urethral orifice* on the ventral surface of the vagina at its junction with the urethra. Near the opening of the urogenital sinus, again in the ventral wall, locate the *clitoris*, the homolog of the penis. On either side of the *urogenital aperture*, at the external opening of the urogenital sinus, just ventral to the anus, are folds of skin called the *labia majora*. These, together with the urogenital aperture, constitute the *vulva*. The prominent, fleshy, conical *urogenital papilla*, which readily identifies the pig as female, projects beyond the urogenital aperture.

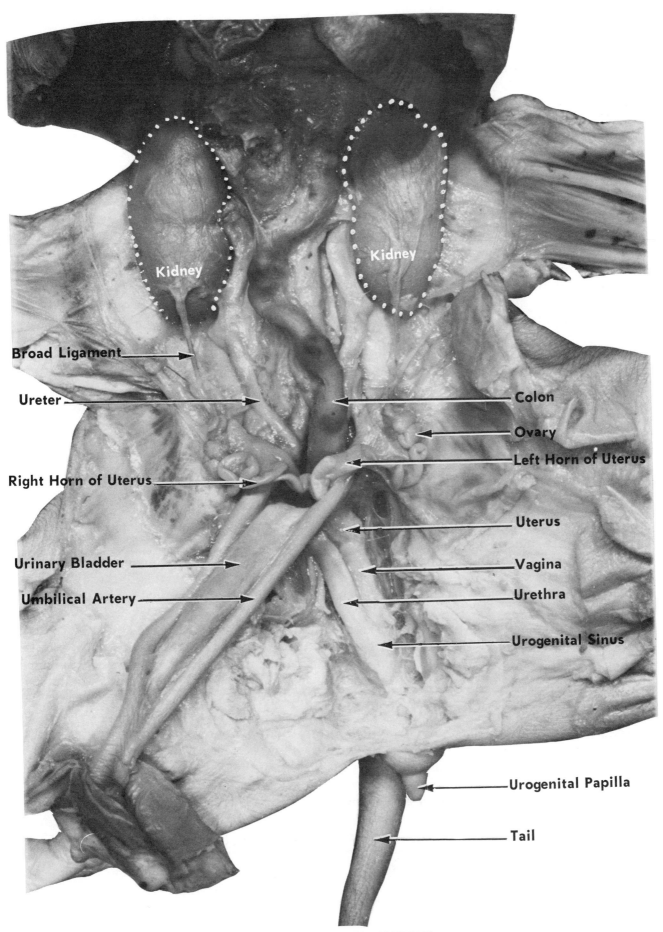

Broad Ligament

Ureter

Right Horn of Uterus

Urinary Bladder

Umbilical Artery

Kidney

Kidney

Colon

Ovary

Left Horn of Uterus

Uterus

Vagina

Urethra

Urogenital Sinus

Urogenital Papilla

Tail

FEMALE UROGENITAL SYSTEM

THE UROGENITAL SYSTEM - MALE

The urinary organs of male pigs, the *kidneys, ureters, urinary bladder,* and *urethra,* are similar to those of the female. Since these have already been described in the last section they will not be repeated here.

The male reproductive structures include the:

Testes — These are the male gonads. Locate the *scrotum,* the swollen double sac ventral to the anus. It contains the *testes.* Carefully cut the skin of the scrotum. It is lined with peritoneum and is divided into two compartments by a medial *septum.*

Epididymis — This is an extremely coiled tubular structure lying on the dorsal surface of each. testes. It consists of a *head* on the anterior part of the testes where it is connected to the testes by numerous microscopic efferent ductules. It also has a *body,* the middle portion, and a posterior portion, the *tail.* Follow the epididymis tail cranially. Its convoluted ducts are continuous with the duct that exits the scrotum into the abdominal cavity.

Vas Deferens (Ductus Deferens) — It is through this tube that sperm and seminal fluid leave the testes. It exits the scrotum into the abdominal cavity.

Note: In order to study the remainder of the male reproductive system cut the pelvic bone at the pubic symphasis and spread the pelvis apart.

Spermatic Cord — The ductus deferens is only one of the tubes leaving the testes. Blood vessels, nerves, and lymphatic vessels supplying the testes also pass from the scrotum into the abdominal cavity. They are united by a tough outer fascia to form the *spermatic cord.*

Inguinal Canal — Follow the spermatic cord cranially through the *external inguinal ring* located at the juncture of the scrotum and the abdominal wall. Continue further cranially through a short channel in the abdominal wall, the *inguinal canal* and out into the abdominal cavity through the *internal inguinal ring.*

During embryological development the testes are at first located within the abdominal cavity, below the kidneys. During later development they descend through the *inguinal canal* into the scrotum. In human males the condition of *inguinal hernia* is common. It is a weakening of the inguinal rings permitting a loop of intestine to be pushed through the inguinal canal into the scrotum. This condition is due to man's upright, two-legged position. Pigs do not suffer from this malady.

Upper Spermatic Cord — Continue to follow the spermatic cord within the abdomen. The blood vessels, the *spermatic vein* and *internal spermatic artery* soon separate from the ductus deferens. The right spermatic vein enters the *posterior vena cava* below the level of the kidney, while the left spermatic vein generally enters the renal vein near the top of the kidney. It thus does not enter the vena cava directly. The right and left internal spermatic arteries enter the *abdominal aorta* next to one another at the level of the posterior region of the kidneys. This can clearly be seen in the photo. Trace these blood vessels in your specimen.

Upper Vas Deferens — The vas deferens loops dorsally over the base of the *ureter* near the urinary bladder and continues caudally. The urethra emerging from the urinary bladder, together with the vas deferens, pass posteriorly and penetrate the *prostate gland* at the proximal end of the *penis.*

108

Urogenital Canal — From this point on, the *urethra* continues as a merged tube, the *urogenital canal*, carrying sperm and seminal emissions from the testes and prostate gland, plus urine from the urinary bladder.

Seminal Vesicles — These are very small glands in the fetal pig and are difficult to locate. They are found dorsally on either side of the prostate gland. They too contribute to the seminal fluid.

Bulbo Urethral Glands (Cowper's Glands) — These are located on either side of the urethra as it passes through the pelvic girdle.

Penis — Follow the urethra, or urogenital canal, caudally to the beginning of the *penis*. The penis is the cylindrical copulatory organ of males. Remove the skin and trace the penis to its attachment in the region of the pubic symphasis. Locate the *crus* of the penis and a small muscle the *ischiocavernosum*. Both are lateral projections at the proximal end of the penis anchored to the ischium bone. The crus is the proximal end of the *corpus cavernosum*, a cylindrical mass of vascular erectile tissue which, together with a second corpus cavernosum lying side by side, form the dorsal part of the penis. A third cylindrical mass of vascular erectile tissue lies ventrally within the penis in a groove between the corpora cavernosum. This is the *corpus spongiosum*. The urethra passes through this mass of vascular tissue. The urethra continues through the penis to its distal end and emerges at the ventral abdominal wall just below the *umbilical cord*. The urethral orifice on the surface of the body is known as the *urogenital opening*. Its presence readily identifies the specimen as a male.

Each student is responsible for learning the reproductive structures of both male and female fetal pigs.

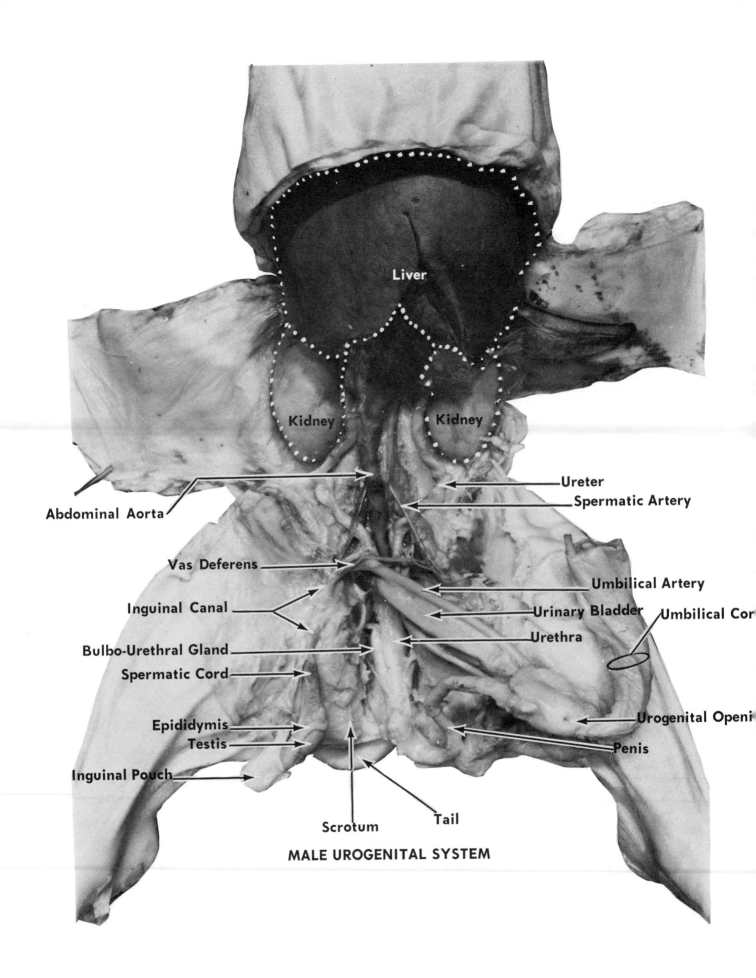

MALE UROGENITAL SYSTEM

Liver

Kidney   Kidney

Abdominal Aorta

Ureter
Spermatic Artery

Vas Deferens

Umbilical Artery

Inguinal Canal

Urinary Bladder   Umbilical Cor

Bulbo-Urethral Gland

Urethra

Spermatic Cord

Epididymis

Urogenital Openi

Testis

Penis

Inguinal Pouch

Tail

Scrotum

SELF - QUIZ VIII
UROGENITAL SYSTEM

1. The human fetus develops in the uterus. In what structure of the mother does the fetal pig develop?
2. The vagina and urethra join to form a single duct in the (a) female pig, (b) female human, (c) both, (d) neither.
3. The urethra and vas deferens join and exit as a single duct in (a) male pig, (b) human male, (c) both, (d) neither.
4. Name the tubes that carry urine from the kidneys to the urinary bladder.
5. Name the blood vessels that lie on either side of the urinary bladder of the fetal pig.
6. The spermatic cord passes into the abdominal cavity through an opening in the abdominal wall known as the _____.
7. Name the duct that carries urine from the urinary bladder to the outside of the body.
8. Egg cells travel from the ovaries through the _____ tubes.
9. Name the blood vessel that supplies the ovary with oxygen and nutrients.
10. Define each of the terms listed below.

ANSWERS

1. _____
2. _____
3. _____
4. _____
5. _____
6. _____
7. _____
8. _____
9. _____
10. a. fallopian tubes _____
    b. seminal vesicle _____
    c. prostate _____
    d. Cowper's gland _____
    e. hernia _____
    f. gynecology _____
    g. hermaphrodite _____
    h. cryptorchidism _____
    i. vasectomy _____
    j. castration _____

Label all of the features of the photograph.

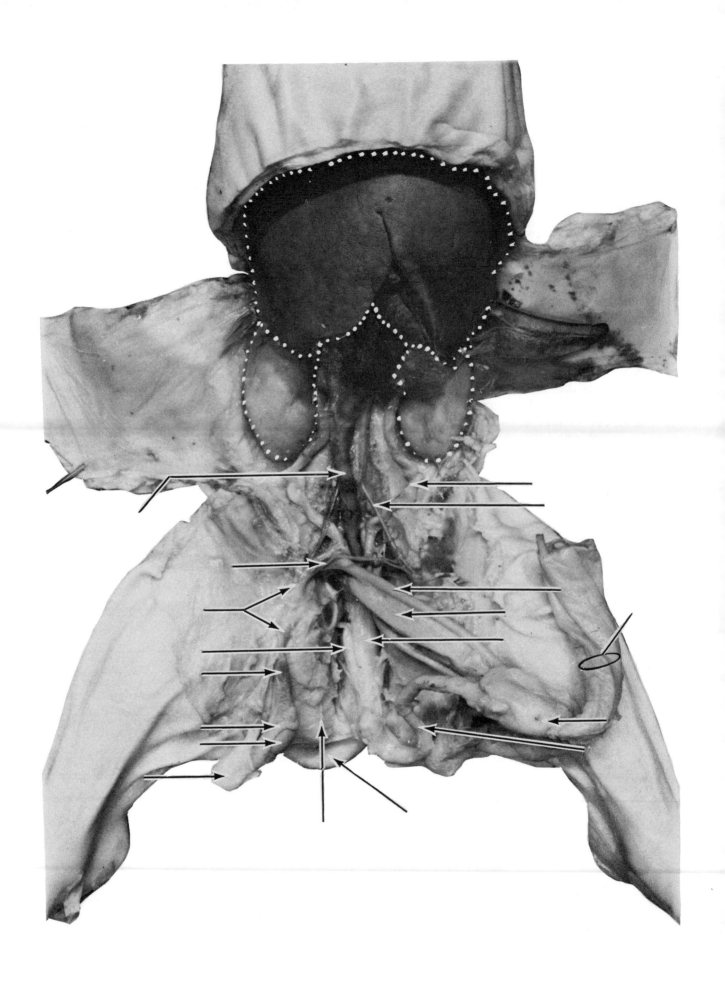

NERVOUS SYSTEM - THE BRAIN

The nervous system is concerned with communications by means of nerve impulses between the parts of the body. It consists of the *central* and *peripheral* nervous systems.

We shall limit our study chiefly to a study of the *brain* and *spinal cord* (central nervous system), plus to one of the specialized sense organs, the *eye*.

To begin, place the fetal pig on the dissection tray with the dorsal surface upward. It is necessary to exercise extreme caution in doing this dissection. It is very easy for the inexperienced novice to injure the delicate brain tissue.

At the top of the skull use your scalpel to cut the skin and the fibrous *epicranial aponeurosis* covering the top of the *cranium* along the mid-dorsal line. Expose the entire cranium. Use a pair of scissors to make the first nicks in the skull bones. Don't penetrate too deeply, the soft brain tissue is easily destroyed. Continue to break and to remove bone fragments with your forceps until the brain is exposed. Bone cutting shears are not needed for the thin and soft bone of the fetal cranium.

Note the *meninges*. These are the series of three membranes covering the brain and spinal cord. The tough fibrous outer *dura mater* adheres to the underside of the cranial bones. The inner, thin, vascular *pia mater* adheres to the surface of the brain and follows its contours and convolutions. The middle layer, the *arachnoid,* will not be seen in these preparations.

The two photos that follow will describe the limits of the brain dissection. The first is a dorsal view, the second a lateral view. The identical specimen is seen in both photos. Note in the lateral view how great a portion of the skull is occupied by the brain. Note the size of the brain in relation to the entire head. We also see the relative dominance, of the *cerebrum* over the *cerebellum* and *medulla oblongata.*

Leave the brain intact and continue to dissect in a posterior direction beyond the brain to expose the spinal cord. After the spinal cord has been dissected we shall return to a detailed study of the sheep brain. At that time you will remove the brain of the fetal pig in order to compare it to the much larger sheep brain.

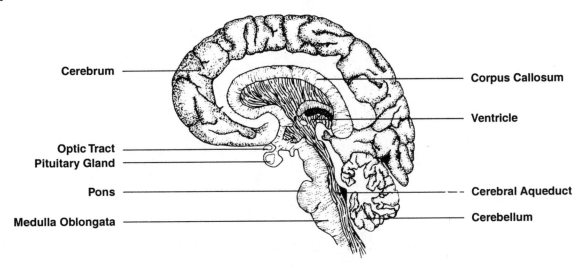

THE HUMAN BRAIN (Sagittal Section)

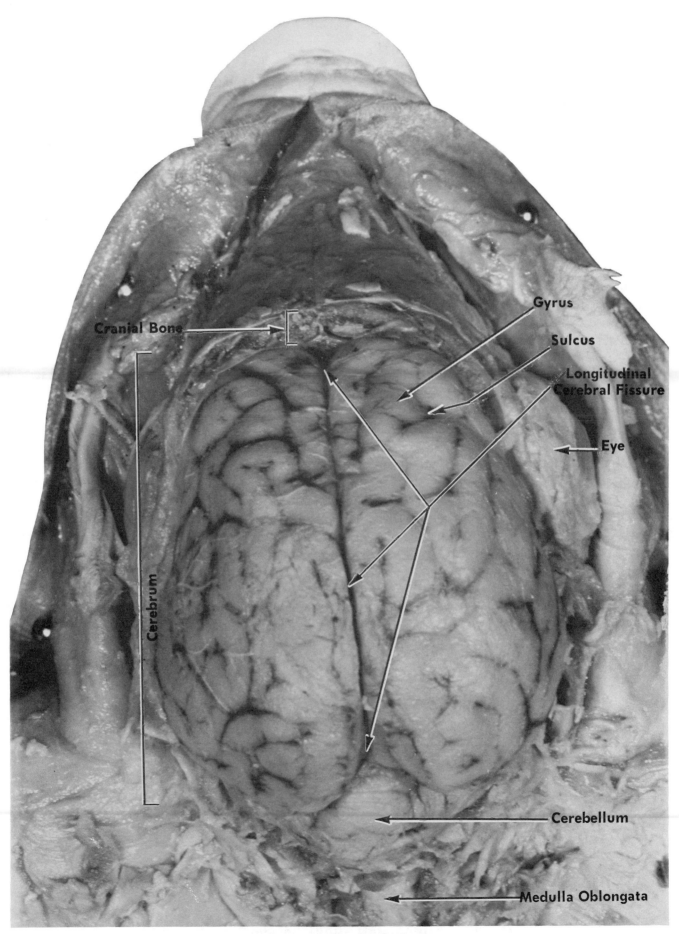

Cranial Bone

Gyrus

Sulcus

Longitudinal
Cerebral Fissure

Eye

Cerebrum

Cerebellum

Medulla Oblongata

BRAIN (Dorsal View)

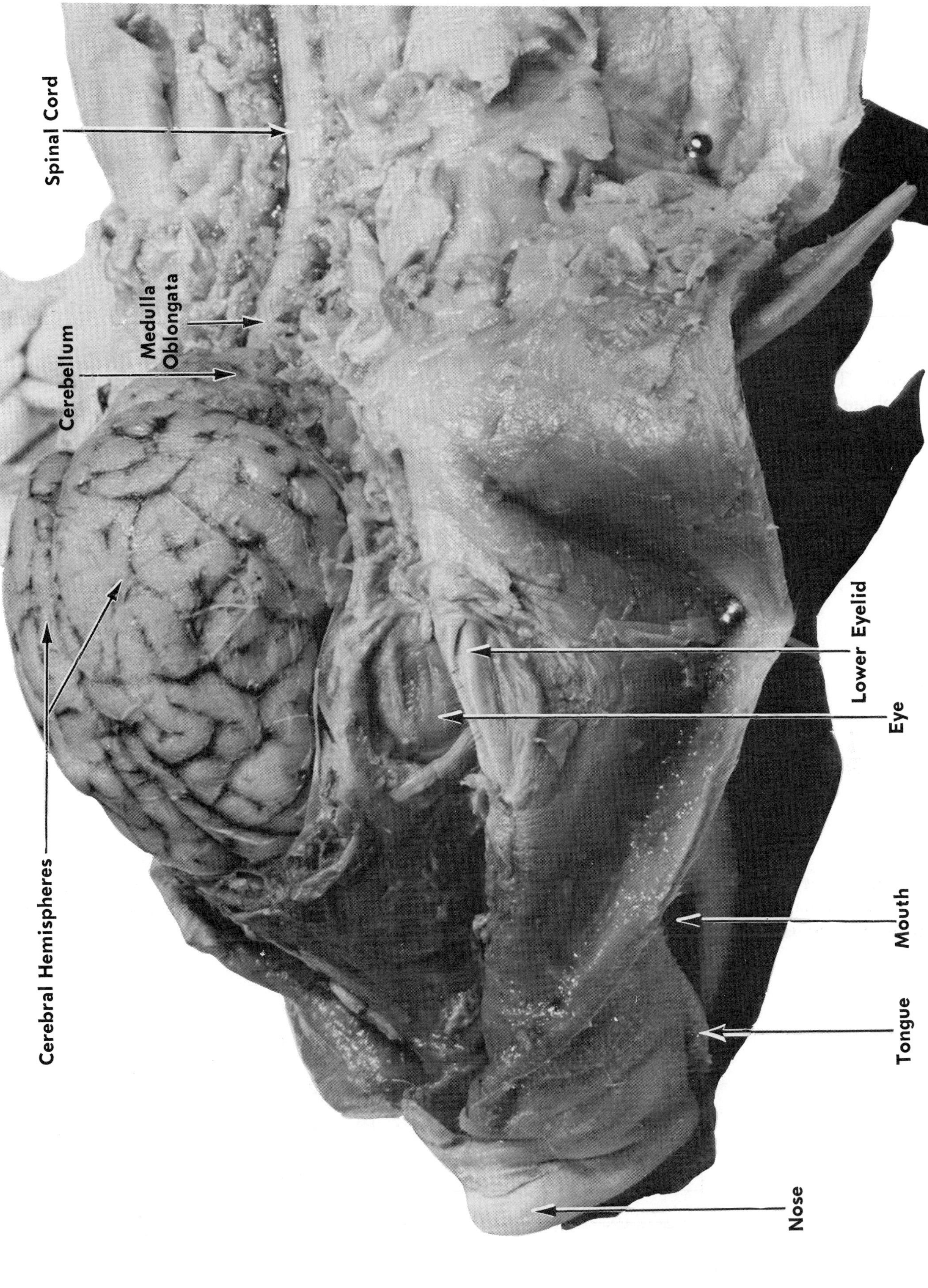

Spinal Cord

Cerebellum

Medulla
Oblongata

Cerebral Hemispheres

Lower Eyelid

Eye

Mouth

Tongue

Nose

BRAIN (Lateral View)

115

NERVOUS SYSTEM - SPINAL CORD

Your task is to expose the entire spinal cord as in the photo, p. 118. The fetal pig is placed on the dissection tray with the dorsal surface upward as for the dissection of the brain.

Begin to expose the spinal cord by first removing the skin and muscles dorsal to the vertebral column. Then using bone clippers and a sharp scalpel cut away the dorsal *neural arches* of the vertebrae to expose the spinal cord along its entire length. Work carefully, one vertebra at a time.

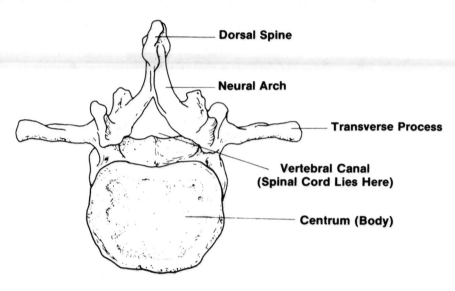

VERTEBRA (Human, Lumbar)

Remove the neural arches from the sides of the vertebrae to expose the origins of the spinal nerves. There are 33 pairs of spinal nerves in the pig, in humans there are only 31. A pair of spinal nerves are given off between each vertebral bone.

Number of Spinal Nerves

	PIG	MAN
CERVICAL	8	8
THORACIC	14	12
LUMBAR	7	5
SACRAL	4	5
COCCYGEAL		1
Total	33	31

Each spinal nerve arises from two roots, the *dorsal root*, which is *sensory*, and the *ventral root*, which is *motor*. These unite a short distance from the cord to carry sensory and motor impulses to and from the spinal cord. For this reason, all spinal nerves are known as *mixed* nerves. Find the prominent rounded swellings on the dorsal root proximal to its union with the ventral root. These enlargements contain the cell bodies of the sensory neurons and are known as *dorsal root ganglia*.

Remove a ¼ inch section of the spinal cord and observe in cross section with a low power dissecting microscope. Locate the *gray matter* in the shape of a capital "H" near the center. It contains the cell bodies of the motor neurons. The white matter around the periphery of the spinal cord is composed of neurons carrying messages up and down the spinal cord, from the brain to muscles and glands. Specialized tracts of white matter communicate with different brain centers.

The *meninges* are membranous coverings for the spinal cord; the *dura mater, arachnoid*, and *pia mater*. They are similar to those described covering the brain.

Note the diameter of the cord along its length. It is thickest in the area of the limbs. The *cervical enlargement* is found near the anterior limbs, while the *lumbo-sacral enlargement* is found near the lower limbs. This corresponds to the many nerves controlling the limbs which originate here.

While the cervical and thoracic spinal nerves exit the spinal column horizontally, those in the lumbar and sacral region pass posteriorly and leave the bony spinal column at a lower region. This causes the appearance of a multi-fibered, tail-like structure within the spinal cord. It makes up the *cauda equina*, literally, horse's tail. The last nerve filament, a remainder of the spinal cord, is called *filum terminal*.

The spinal nerves to the limbs form complex networks, interconnected one to the other, known as plexuses. The anterior *brachial plexus* to the forelimbs is formed by the union of the branches of the last three cervical and the first thoracic nerves. The posterior *lumbosacral plexus* is composed of the branches of the last three lumbar and first sacral nerves.

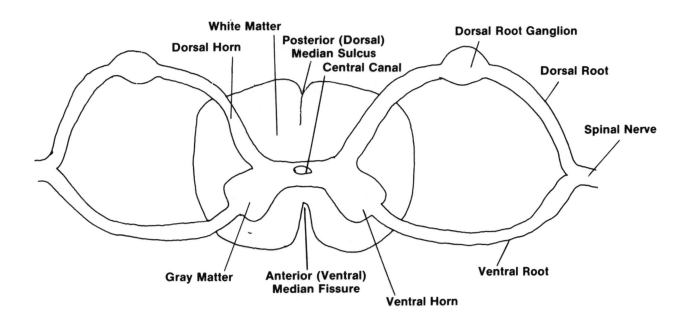

HUMAN SPINAL CORD (Cross Section)

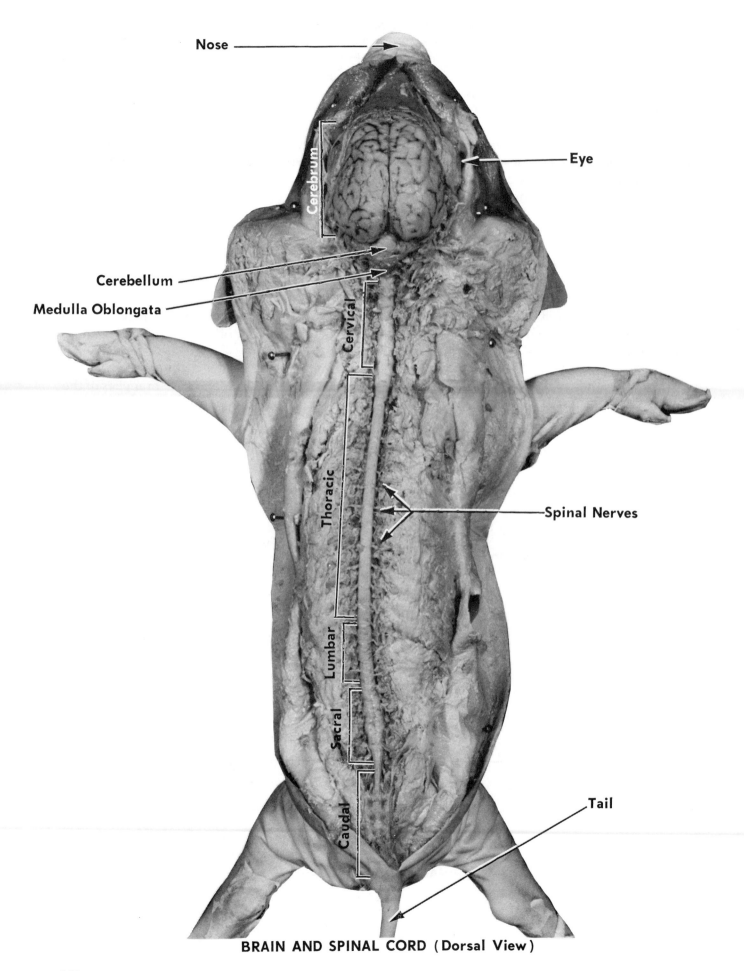

Nose

Eye

Cerebrum

Cerebellum

Medulla Oblongata

Cervical

Thoracic

Spinal Nerves

Lumbar

Sacral

Tail

Caudal

BRAIN AND SPINAL CORD (Dorsal View)

SPINAL NERVES

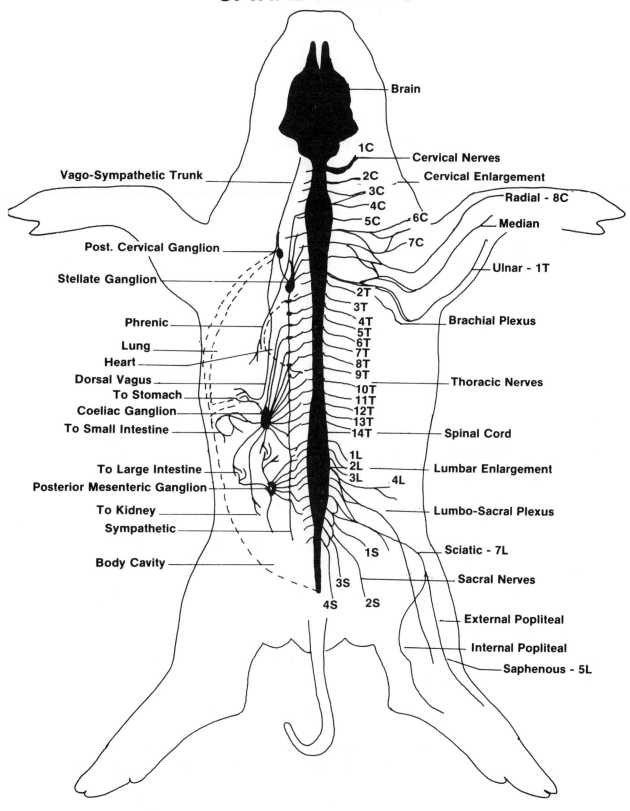

Brain

1C — Cervical Nerves

Vago-Sympathetic Trunk

2C — Cervical Enlargement

3C
4C
5C
6C
7C

Radial - 8C

Median

Post. Cervical Ganglion

Ulnar - 1T

Stellate Ganglion

2T
3T
4T
5T
6T
7T
8T
9T

Brachial Plexus

Phrenic

Lung

Heart

Dorsal Vagus

To Stomach

Coeliac Ganglion

To Small Intestine

Thoracic Nerves

10T
11T
12T
13T
14T

Spinal Cord

1L
2L
3L
4L

To Large Intestine

Posterior Mesenteric Ganglion

Lumbar Enlargement

To Kidney

Sympathetic

Lumbo-Sacral Plexus

1S

Sciatic - 7L

Body Cavity

Sacral Nerves

3S
4S
2S

External Popliteal

Internal Popliteal

Saphenous - 5L

SHEEP BRAIN

The sheep brain is very similar to that of the pig and human. Its large size makes it particularly well suited for study. Structures too small to be noticed in the fetal pig brain may be readily identified in the sheep brain. Remove the exposed brain of the fetal pig and compare it to that of the sheep throughout this exercise.

Dorsal View — Remove any portion of the *dura mater* still adhering to the surface of the brain.

Note the large *cerebral hemispheres* and their *convoluted* surfaces. The *sulci* are depressions and the *gyri* are raised areas of the surface. The right and left cerebral hemispheres meet at the *longitudinal cerebral fissure.*

You may easily identify the three major regions of the brain, the *cerebrum, cerebellum* and *medulla oblongata*.

Ventral View — More detailed structures are seen upon the ventral surface (underside) of the brain.

The sensory tract from the eye, the *optic nerve*, is seen in the photo, p. 123. Most of the fibers of this nerve cross to opposite sides of the brain thereby forming the "X" shaped *optic chiasma* clearly visible in the photo.

Terminals of other cranial nerves are also visible.

Note:

— The *olfactory bulb* at the anterior end of the cerebrum.

— The thickest of all of the cranial nerves, the *trigeminal nerve*, originates at the anterior lateral surface of the *pons.*

— *The vagus nerve* is the longest of the cranial nerves. It exists from the lateral surface of the *medulla oblongata* and sends branches to the pharynx, larynx, heart, lungs, stomach and the intestines.

The brain is well supplied with blood. In the photo, p. 123, we see some of the arteries that join at the base of the brain (anastomosis) to form the *Circle of Willis*. Learn the names of these arteries.

The pituitary gland, or *hypophysis*, has been removed in this specimen. This often happens when the brain is forcibly removed from the cranium.

Sagittal View — Make a mid-sagittal section of the sheep brain. This view shows more details than the two seen previously. See photo, p. 124.

Note:

— The *Arbor Vitae* (Tree of Life) — This is the branched white matter within the *cerebellum*.

— The *Right Lateral Ventricle* — It can be seen bounded by the *corpus callosum* above and the *fornix* below. The corpus callosum is located at the base of the longitudinal cerebral fissure and connects the right and left cerebral hemispheres.

— The *Pineal Body*.

— The *Cerebral Aquaduct* — This passageway joined the third to the fourth ventricle.

— The *Fourth Ventricle*.

— The *Thalamus* (intermediate mass).

— The *Corpus Callosum* — This is a band of white fibrous tissue connecting the right and left halves of the brain. It forms a roof over the large *lateral ventricles* of the brain in which *cerebrospinal fluid* is found.

Locate and identify all of these structures in your specimen.

Coronal Section — Cut the brain in coronal section, perpendicular to the *longitudinal cerebral fissure*. Observe the difference in coloration between the outer ⅛ inch of brain tissue and that beneath. The darker tissue, or *gray matter,* is the *cortex.* Note that as a result of the extensive convoluted pattern of the cerebrum, the "gray" cortex extends deeply into the surface. The cortex contains the *cell bodies* of the brain's *neurons* while the white matter is composed of *myelinated axons.* Two thirds of the nerve cells of the entire body are located in the thin outer layer, the cortex, of the brain. Note too, the reversal of positions of gray and white matter between the brain and spinal cord. While the gray matter is toward the outer surface of the brain, it is centrally located in the spinal cord, surrounded by white matter (see p. 117).

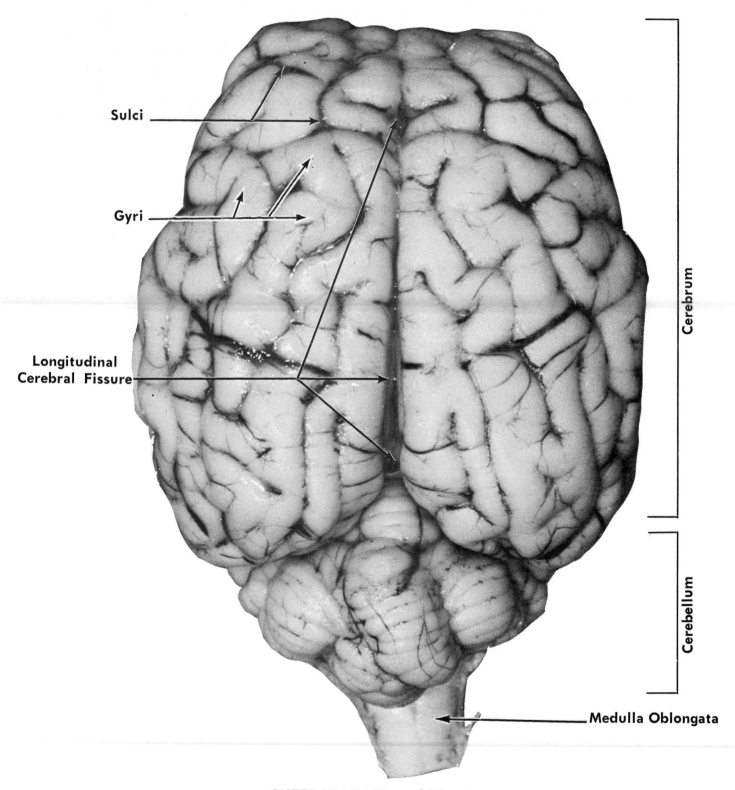

Sulci

Gyri

Longitudinal
Cerebral Fissure

Cerebrum

Cerebellum

Medulla Oblongata

SHEEP BRAIN (Dorsal View)

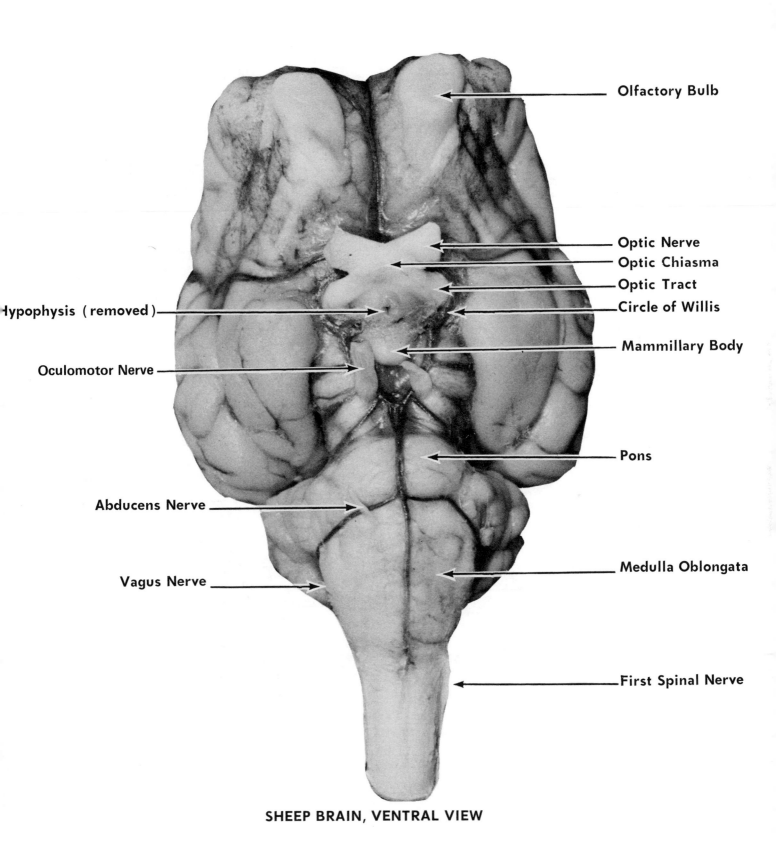

Olfactory Bulb

Optic Nerve

Optic Chiasma

Optic Tract

Circle of Willis

Hypophysis (removed)

Mammillary Body

Oculomotor Nerve

Pons

Abducens Nerve

Medulla Oblongata

Vagus Nerve

First Spinal Nerve

SHEEP BRAIN, VENTRAL VIEW

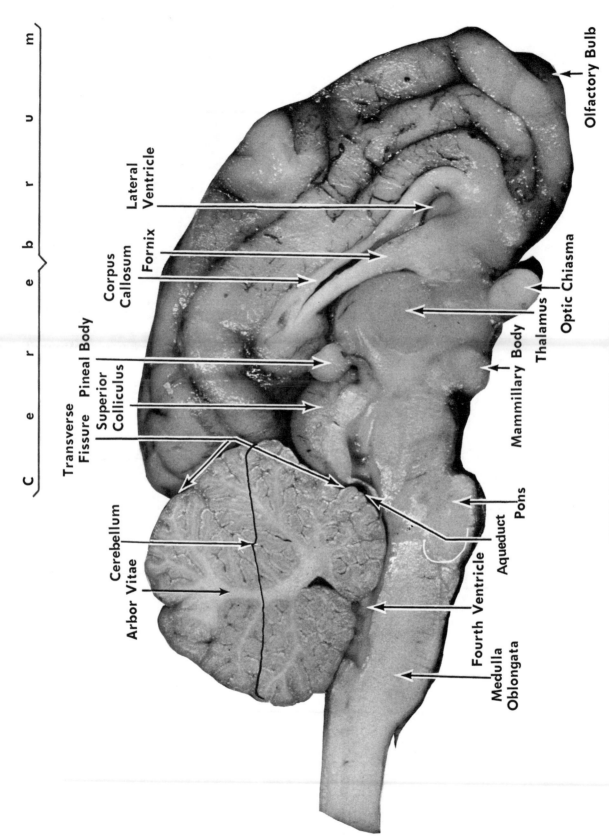

Cerebrum

Lateral
Ventricle

Corpus
Callosum

Fornix

Transverse Pineal Body
Fissure

Superior
Colliculus

Cerebellum

Arbor Vitae

Olfactory Bulb

Optic Chiasma

Mammillary Body

Thalamus

Pons

Aqueduct

Fourth Ventricle

Medulla
Oblongata

SHEEP BRAIN, SAGITTAL VIEW

124

SELF - QUIZ IX
BRAIN AND SPINAL CORD

1. The three major divisions of the brain are the a) _____, b) _____, and the c) _____.
2. The folds upon the surface of the brain are known as _____
3. The two hemispheres of the cerebrum are separated by the _____ fissure.
4. The optic nerves from the eyes cross before entering the brain. The crossed structure is known as the _____.
5. From what part of the brain does the vagus nerve originate?
6. The hypophysis is more commonly known as the _____.
7. a. How many ventricles are there in the brain?
   b. What is contained within the ventricles of the brain?
8. The brain is protected by a portion of the skull known as the _____.
9. The swellings on the dorsal roots of spinal nerves are known as _____.
10. Define each of the terms listed below.

ANSWERS

1. (a) _____
   (b) _____
   (c) _____
2. _____
3. _____
4. _____
5. _____
6. _____
7. a. _____
   b. _____
8. _____
9. _____
10. a. meninges _____
    b. olfactory bulb _____
    c. cranial nerve _____
    d. corpus callosum _____
    e. fornix _____
    f. cortex _____
    g. sulci _____
    h. gyri _____
    i. ganglion _____
    j. Circle of Willis _____

Label all of the features of the photograph.

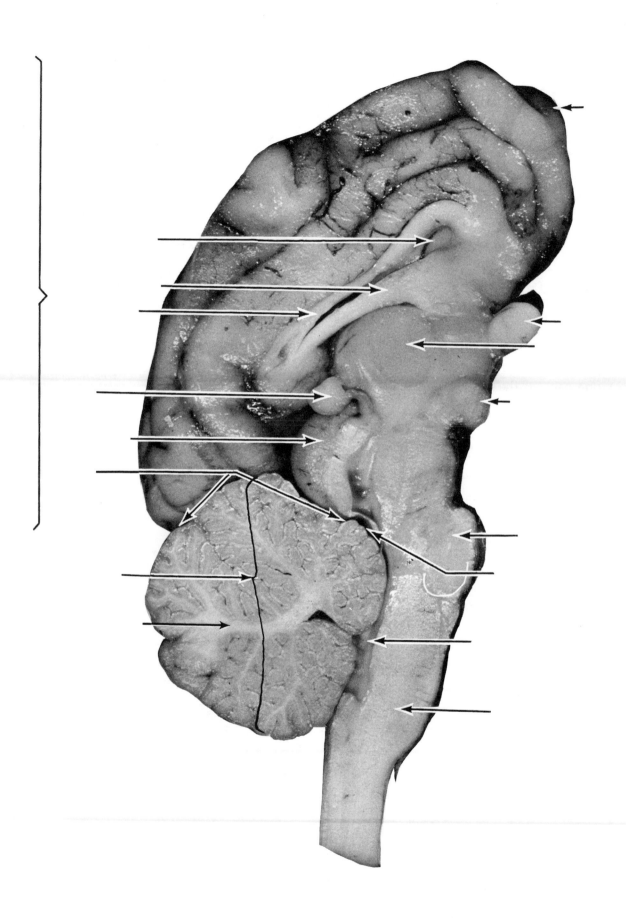

SHEEP EYE - EXTERNAL VIEW AND TRANSVERSE SECTION

The external view of the eye reveals some of the extrinsic *eye muscles* which move the eye. Also seen is the thick layer of *fat* covering the rear of the *eyeball*.

The transverse section reveals the interior of the eyeball. The fat and muscle layers have been removed. The stump of the *optic nerve* may be seen exiting from the rear of the eye.

The *sclera, choroid* and *retina* are seen.

Locate and identify on your specimen all of the structures pictured.

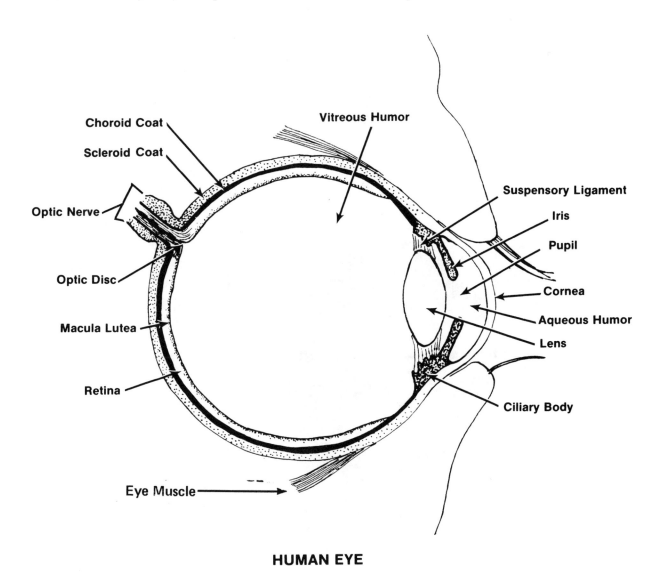

HUMAN EYE

External View

SHEEP EYE

Transverse Section

Cornea
Anterior Chamber
Iris
Lens
Posterior Chamber
Sclera
Optic Nerve
Pupil
Choroid
Retina

Cornea
Sclera
Eye Muscle
Fat
Eye Muscle